チェルノブイリ 春

写真・文 中筋 純

The Revelation of Chernobyl
by Nakasuji Jun

二見書房

まえがき

　2007年の晩秋にチェルノブイリを初めて訪問した。草木も生えぬ死の風景を想像していたが、目の前に広がったのは赤や黄色の宝石がちりばめられたような紅葉の大地だった。眠りにつく自然の営みの最後の響宴は華やかでどこかはかない。時折吹く白ロシアからの乾いた北風に木々の華やかな衣装は一枚一枚はがされ、カサカサと乾いた音を立て吹き寄せられるように廃墟のコンクリートアパートの玄関口に集まってゆく。長い冬はすぐそこまで来ているような気がした。森羅万象の生生流転が世界中で平等に繰り広げられるとしても、目の前の核汚染された大地に半年後には木々が萌え、小鳥たちのラブソングが再び聞こえるのだろうか？　そんな思いの中、前作『廃墟チェルノブイリ』を出版した。

　2009年5月。僕はキヤノンサロンで催した「黙示録チェルノブイリ」の個展期間の隙間を縫うようにしてこの地への再訪を果たした。出迎えてくれたのは前回と同じスタッフたち。通訳のアレクセイは体調を崩したこともあってか前より幾分か痩せたようだが翻訳マシン？　のようなしゃべり口は健在、ガイドのサーシャは相変わらずウォッカが切れると機嫌が悪くなり、ドライバーのセルゲイは念願だったルノーの新車を手に入れた。

　チェルノブイリに向かう車の窓から吹き込む風は暖かかった。鉄条網に囲まれた草地も若草に覆われ、木々はすっかりと緑の衣装に衣替えをした。小鳥たちは相手を求め至る所で喉を自慢して森は艶っぽいメロディーに埋め尽くされている。街道の路傍にはタンポポが咲き乱れ、イノシシはウリ坊を率いて食料探しに余念がない。季節は移ろい、春の沸き立つ生命の息吹がチェルノブイリを漂っているのだ。世界に等しく訪れるはずの春なのにこれほどいとおしく感じたことはなかった。

　再訪の喜びは一瞬放射能の存在をぬぐい去ってしまった。森の中にはいまだに放射線警告標識の黄色い看板が野に咲く毒花のように睨みを効かせ、発電所4号炉の石棺付近ではいまだにガイガーカウンターが無機質な電子音を響かせる。放射能が姿を変え移り変わることもなくこの地に存在しているのはどうやら紛れもない事実のようだ。1986年の事故から今年で25年が経つというのにだ。

　発電所城下町のプリピアチでは薄緑の衣装をまとった街路樹のポプラが大きく枝を張り立派な森を作り上げていた。木々が発する青臭い樹液の匂いが漂い、以前撮影で行った新緑の白神山地のブナの森を彷徨っているような錯覚に陥ってしまった。命あふれる春の森だ。

「純さん、新しい場所に行ってみましょうよ」

　アレクセイが指差したのは街の北端にある高層マンションだった。

「春のチェルノブイリが見えますよ」

　僕たちは息を切らしながら瓦礫に埋もれた薄暗く狭いマンションの階段を一歩一歩登っていった。春のチェルノブイリを見るために。

Preface

I visited Chernobyl for the first time in late autumn four years ago. Before actually going there, I imagined the town should have the scenery of death where no plants and trees grew. But what I saw instead was a land full of beautiful autumn leaves, which looked like red and yellow jewelry. The beauty was the final festival of nature before it slept in winter. It was somehow frail. Dry and cold winds intermittedly blew in from White Russia, stripping away the gorgeous dresses of the trees. The leaves drifted off with rustling sounds to the entrances of abandoned concrete apartments. The scene made me think a long winter was coming soon.

Winter is followed by spring. Although all creatures in the universe equally repeat their life and death, I wondered whether the trees in the land contaminated by radiation put forth buds in spring and birds sang love songs. Bearing the doubt in mind, I published the previous book titled ``Revelations Chernobyl.''

In May 2009, I managed to visit the city again in the short period of time between my photo exhibitions titled ``Apocalypse Chernobyl'' at Canon Salon in Tokyo. Those who welcomed me were the same staff who had helped me during my first visit to the city. Alexei, the interpreter, looked like he had lost weight because of sickness; but, he talked fast like a translation machine in the same way as before. The local guide, Sasha, still had his bad moods whenever he got sober after drinking vodka. Sergei, the driver, bought a new Renault, something he had wanted to own for a long time.

On the way to Chernobyl, warm winds blew into the car. Grass fields enclosed by barbed wire fence were covered by young grass. Trees grew thick with green leaves. Birds were proudly singing to seek their mates and the forests were filled with romantic melodies. Dandelions were blooming on the roadsides and wild pigs were leading their children to look for food. The nature showed the season had changed and the land ushered the new vivid spirit of spring. Spring comes everywhere in the world. But I had never cherished spring so much before visiting Chernobyl at that time.

The joy got rid of the existence of radiation for a moment. But yellow signs of radiation stood as warnings in the forests, the signs looked like poisonous flowers. In the area of the No. 4 reactor enclosed in a concrete and lead sarcophagus, Geiger counters still detected radiation with their noisy electric sounds. Apparently it confirms the fact that radiation has continued to exist in this land, though 25 years have passed since the catastrophe in 1986.

In Pripyat, the town where the nuclear power plant is located, poplars with pale green leaves on the streets grew high and formed large forest. It smelled of the fresh saps of the trees, which provided me the illusion that I was walking in the forest of beech trees in Shirakami Mountain Range in Northern Japan where I had once visited for my job.

``Jun, let's go to the place where we've never been,'' Alexei said, pointing out a high-rise condominium in the northern edge of the town. ``We can see the spring scenery of Chernobyl.''

We went up the narrow stairs of the dark condominium step by step, losing our breath, to see Chernobyl in spring.

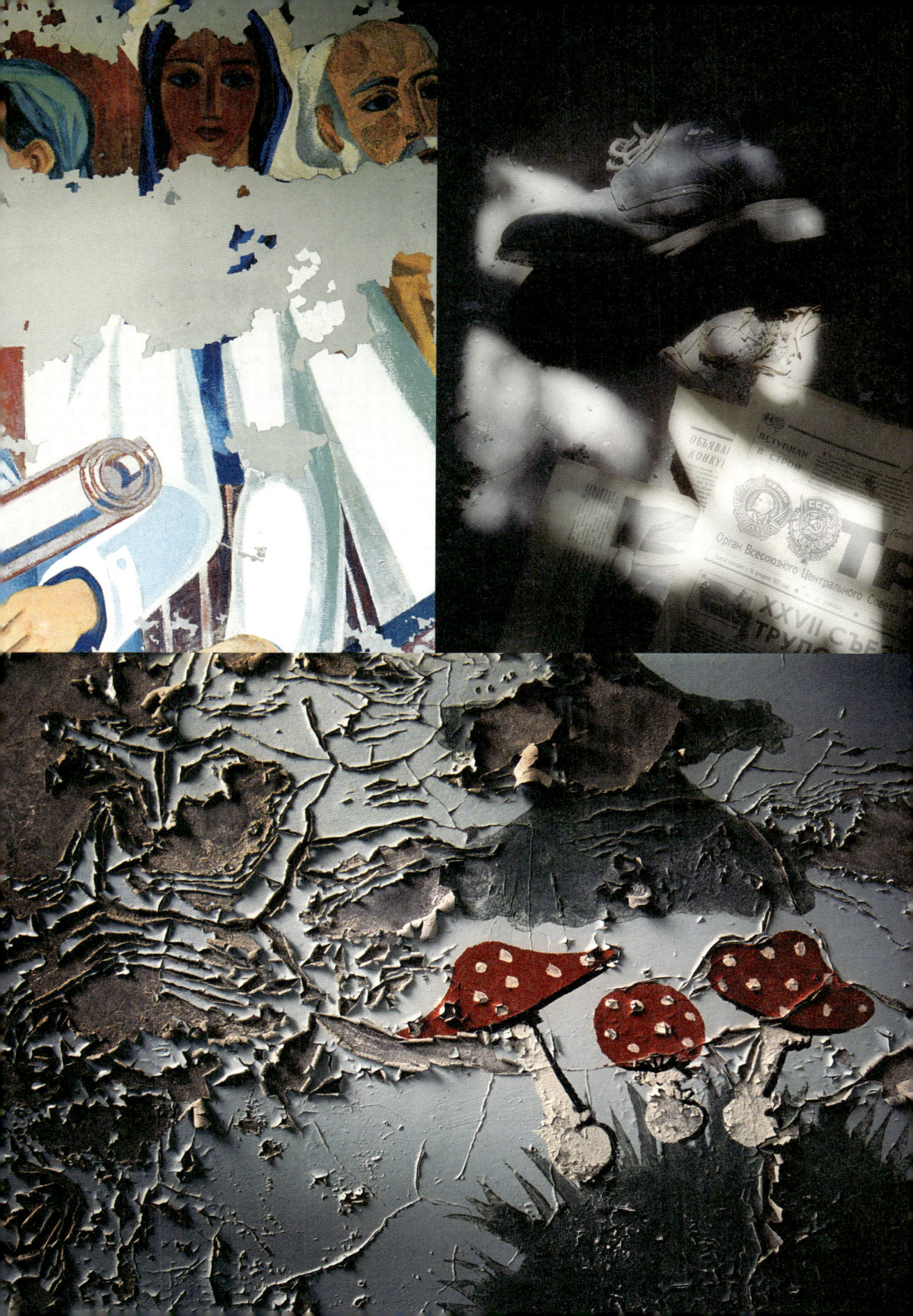

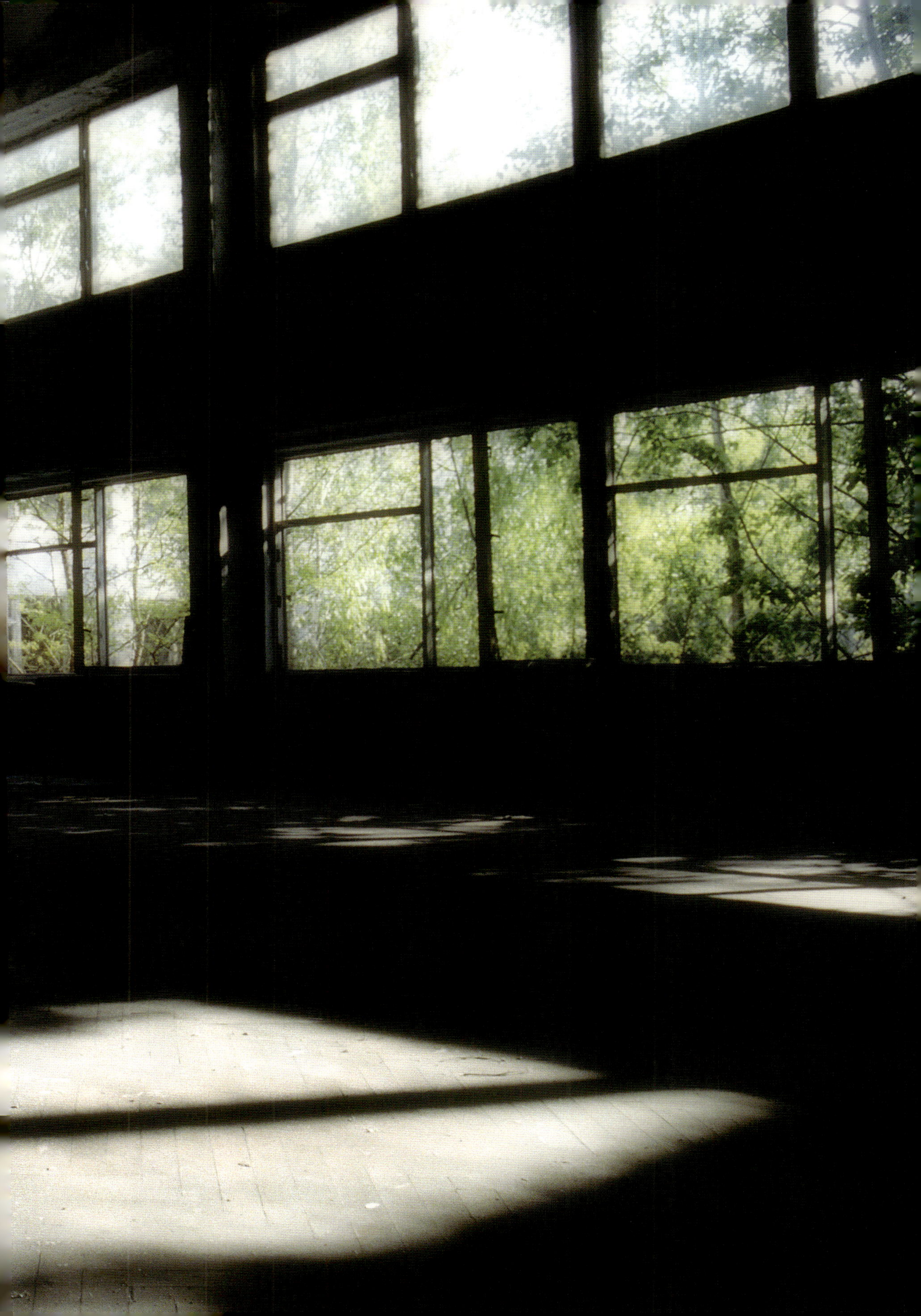

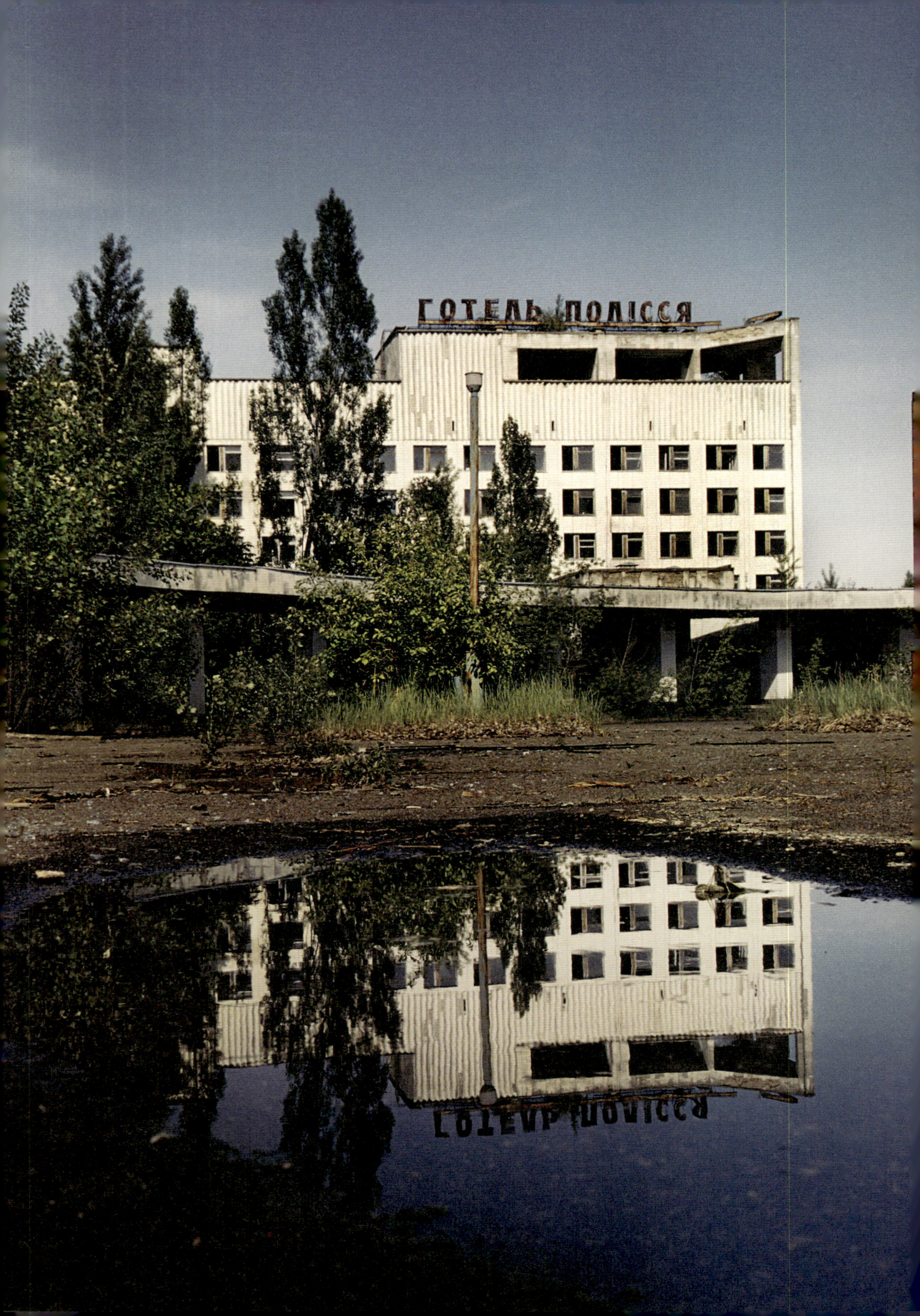
ГОТЕЛЬ ПОЛІССЯ

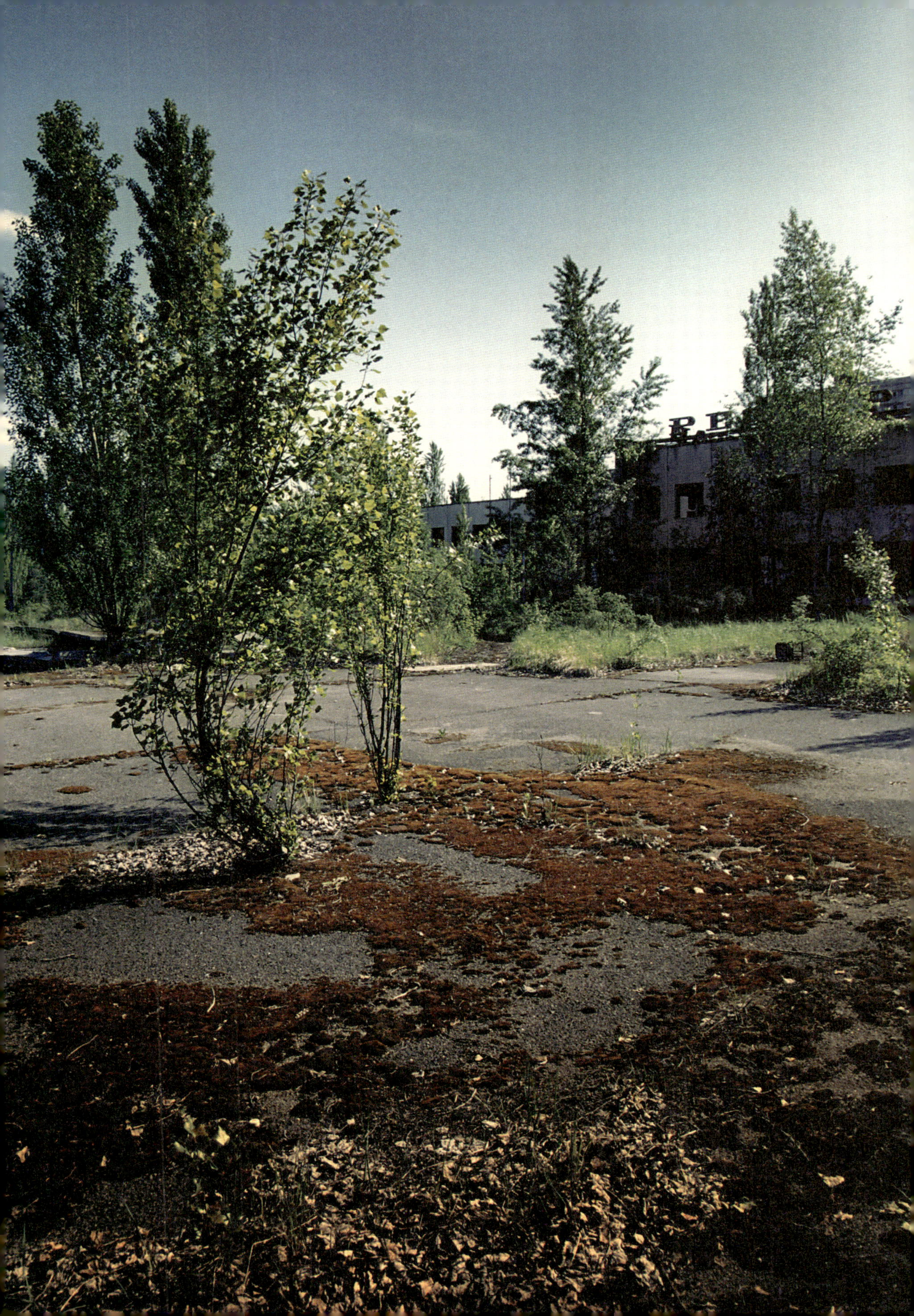

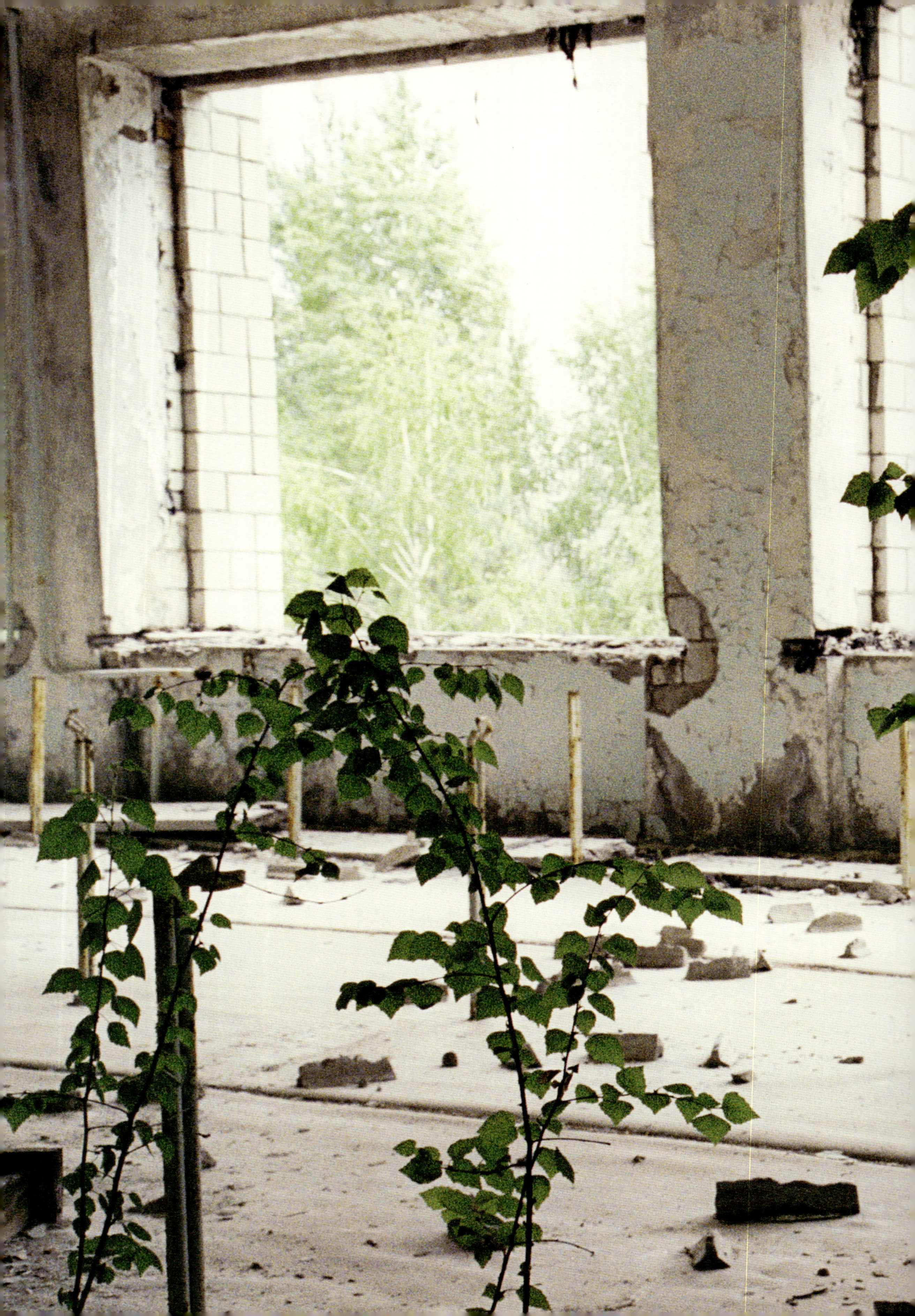

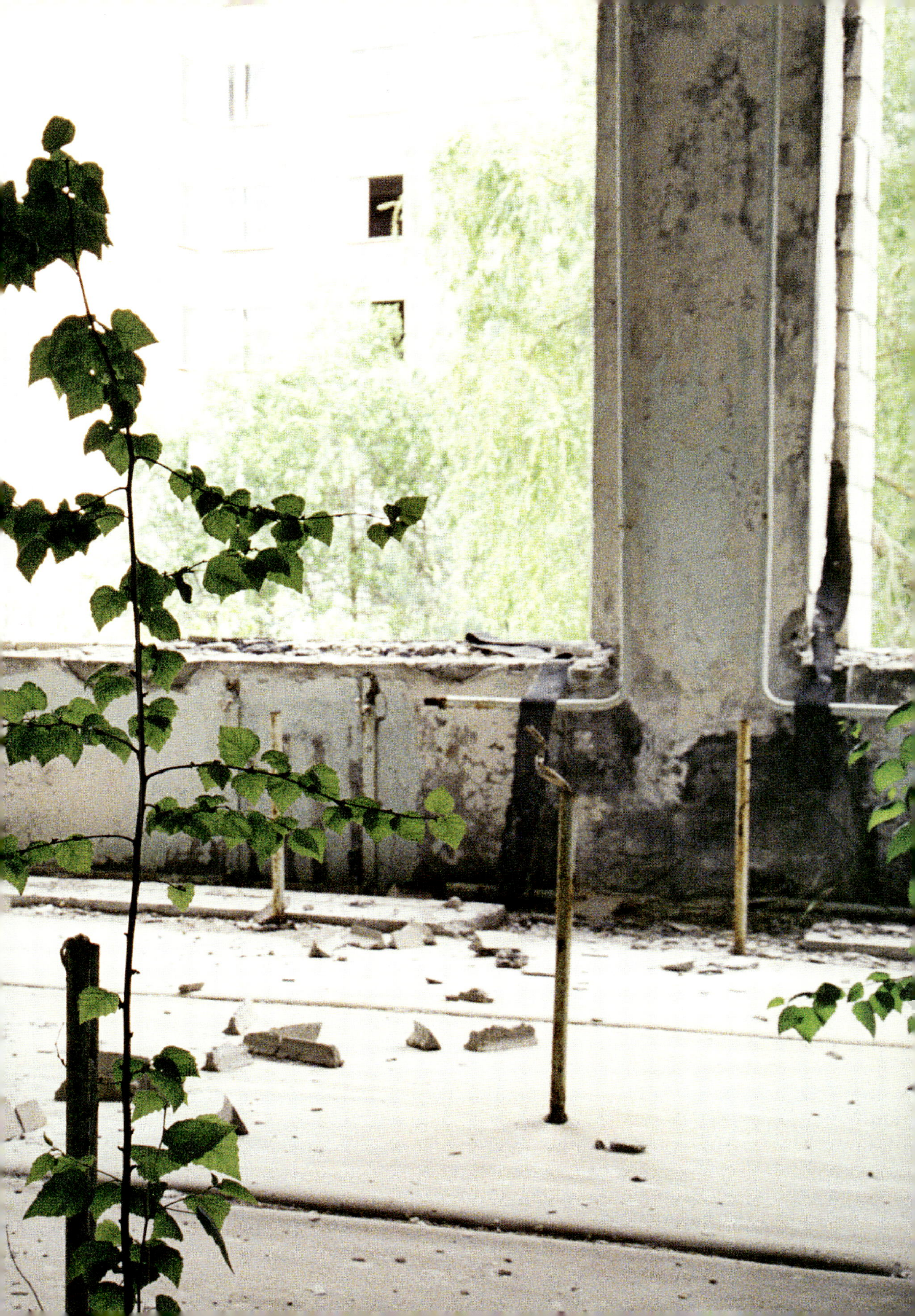

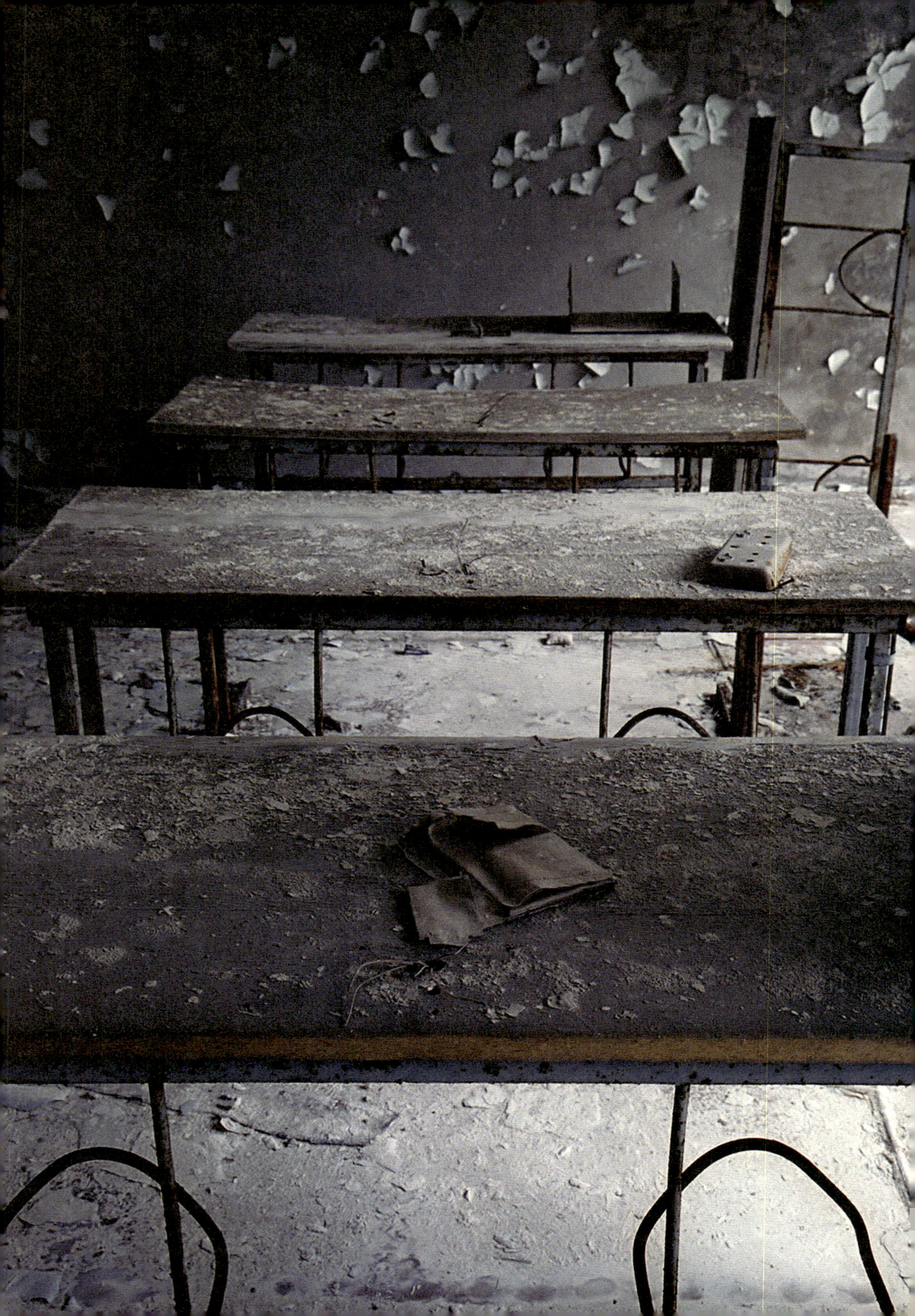

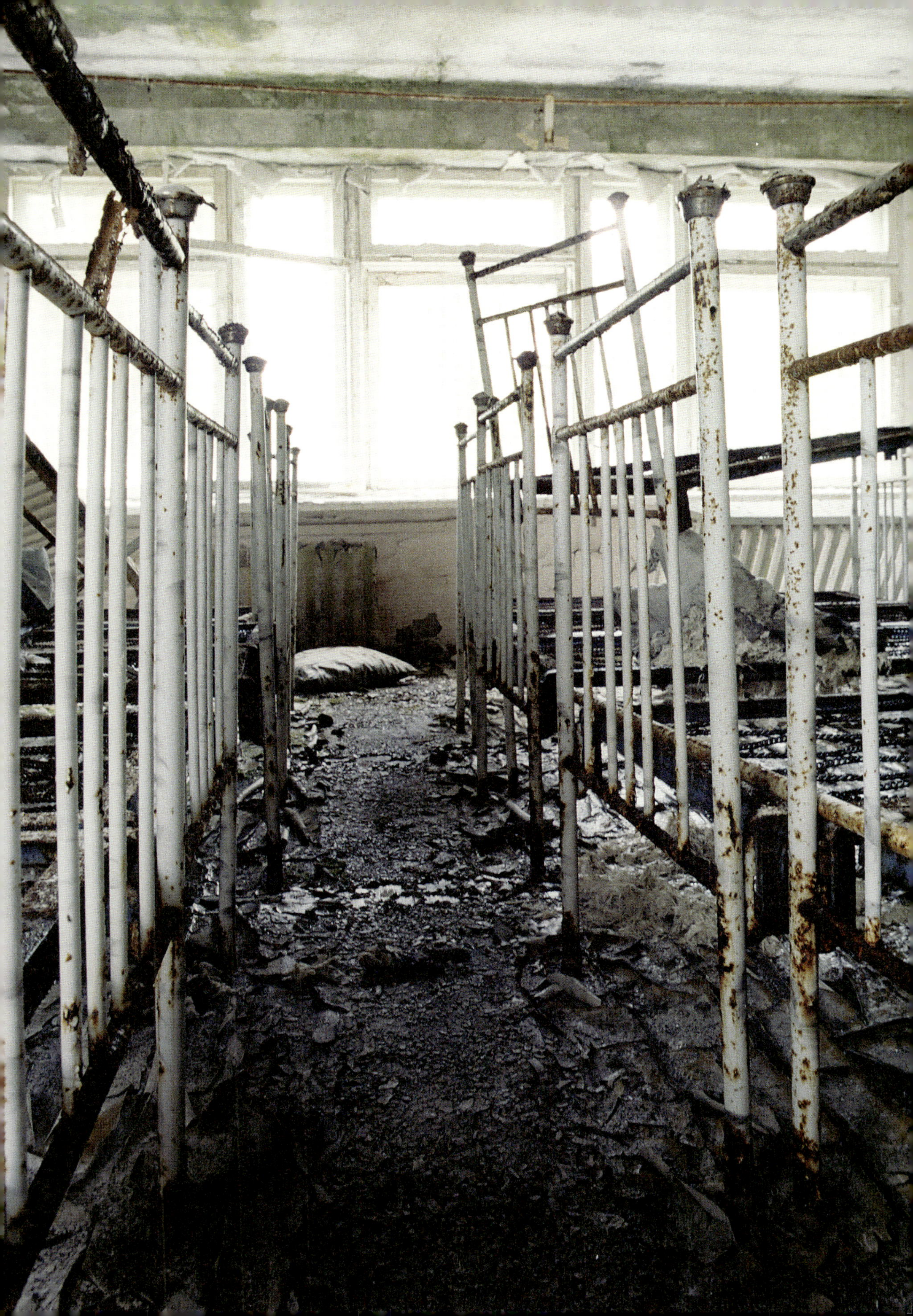

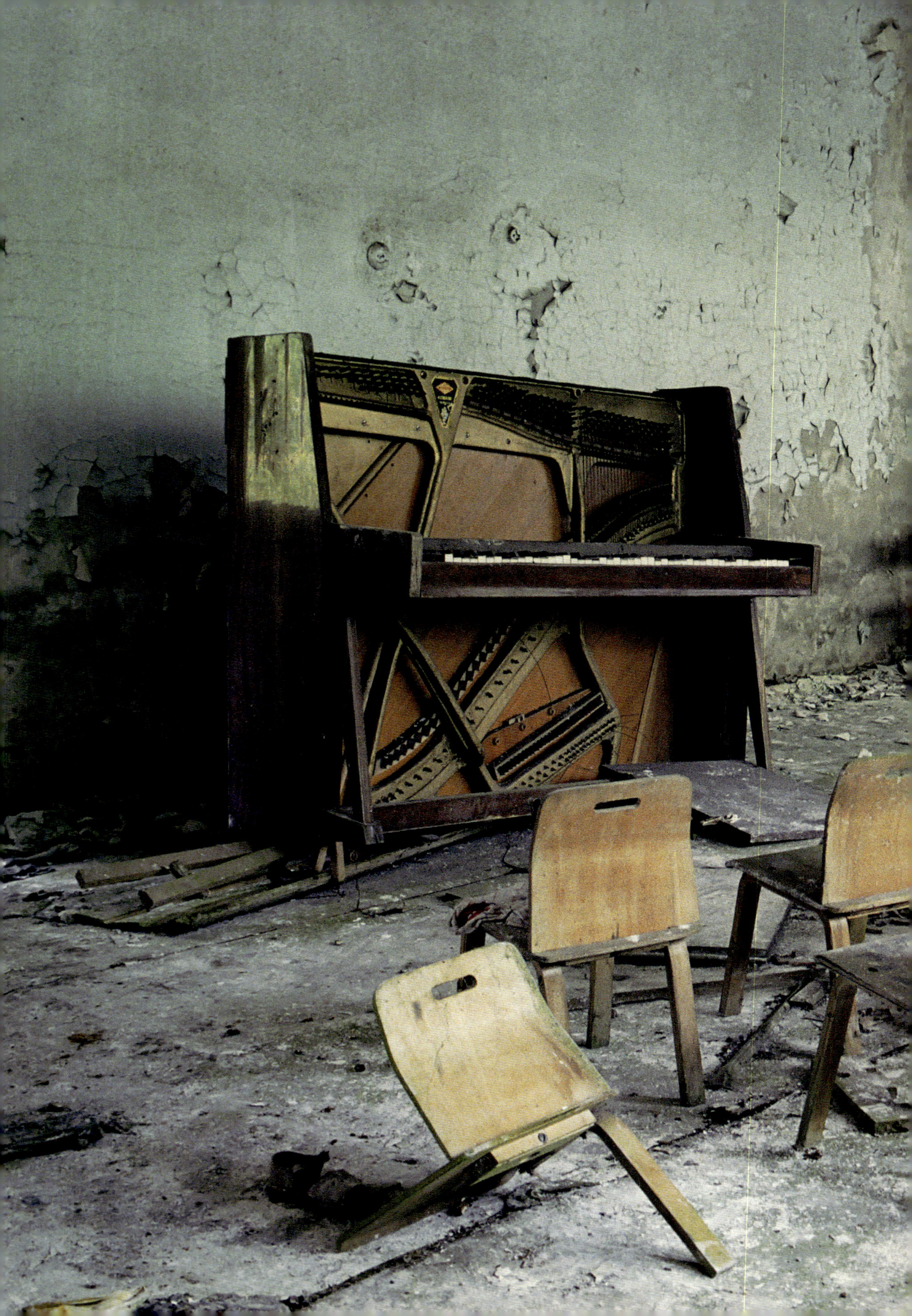

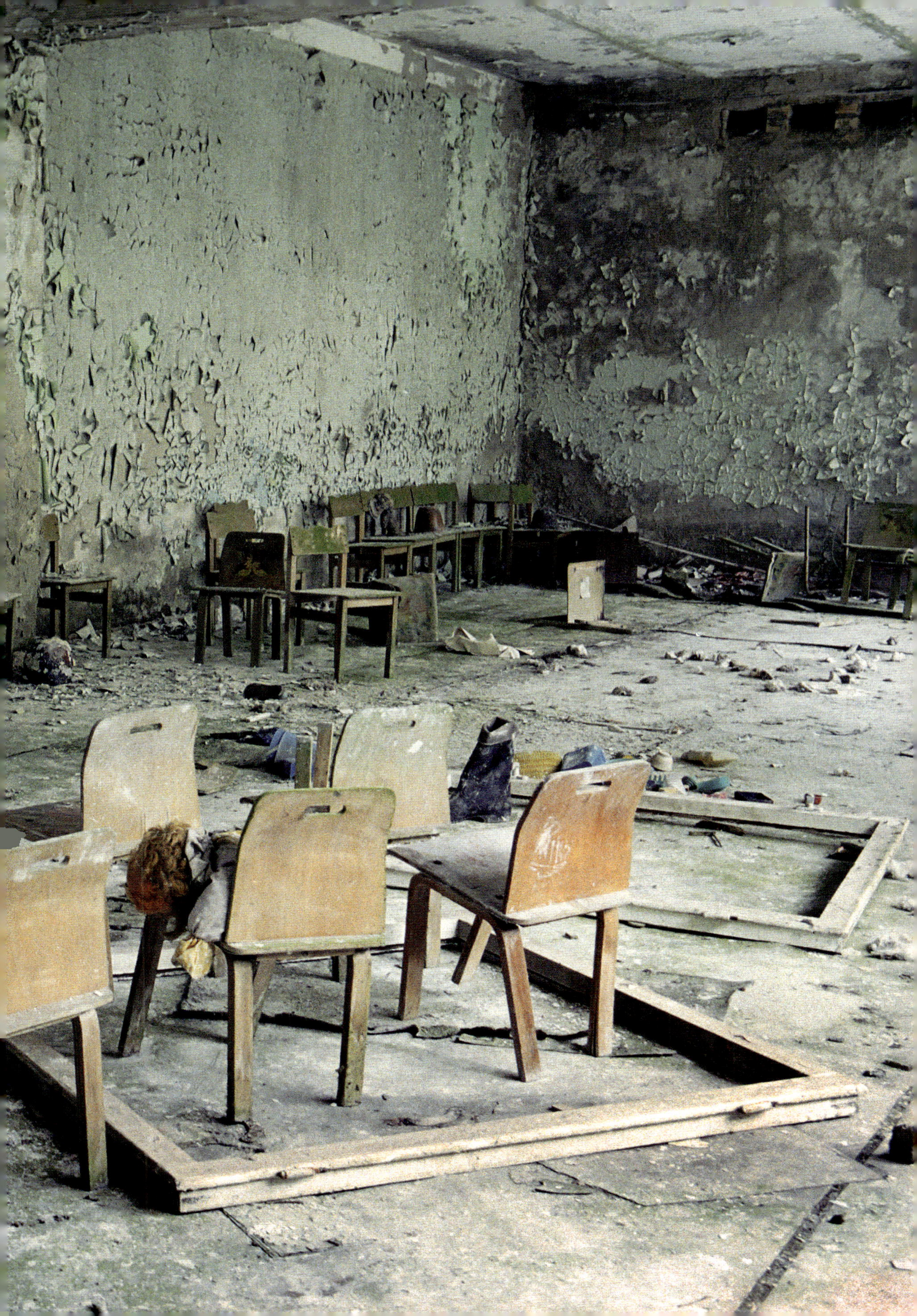

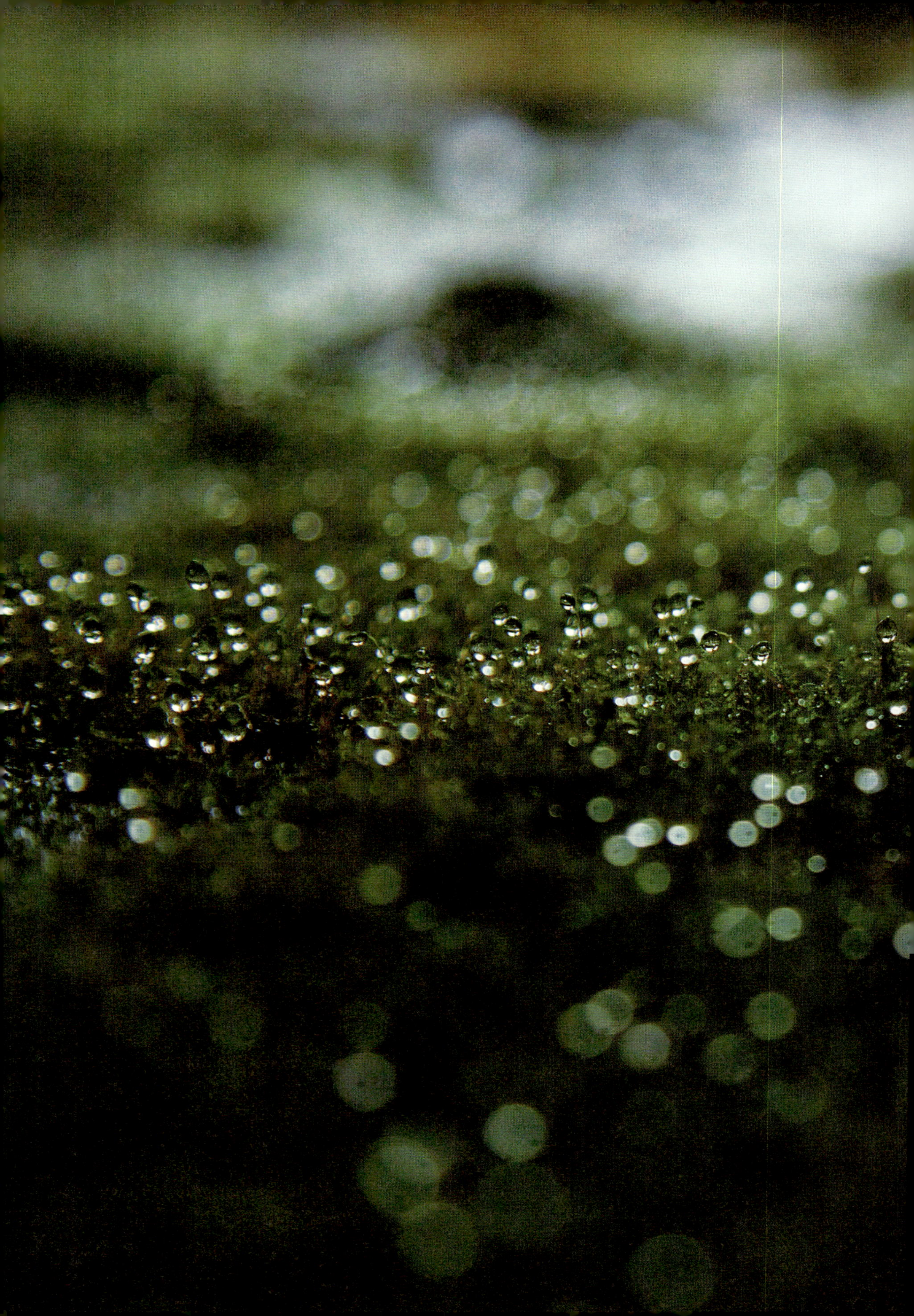

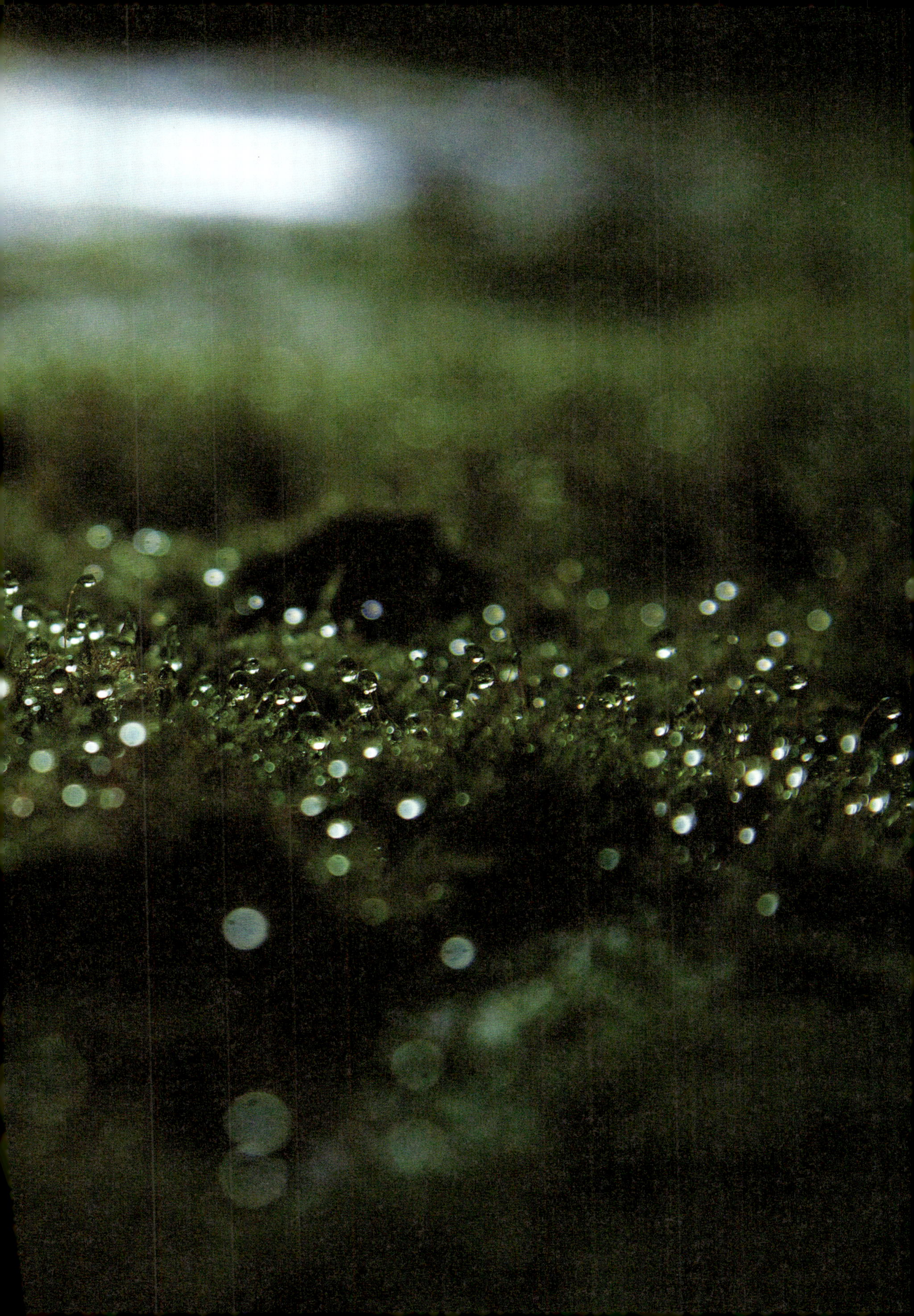

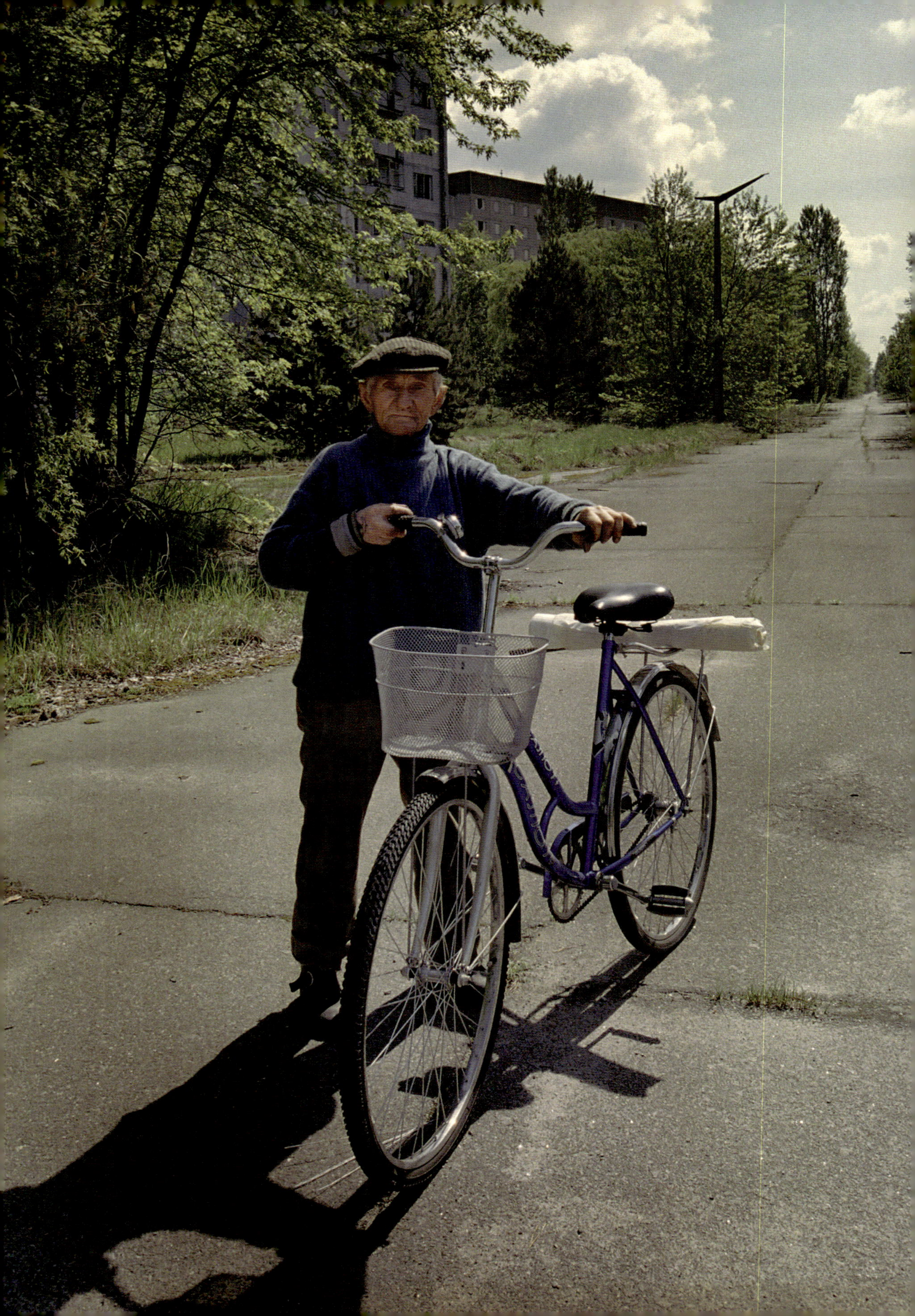

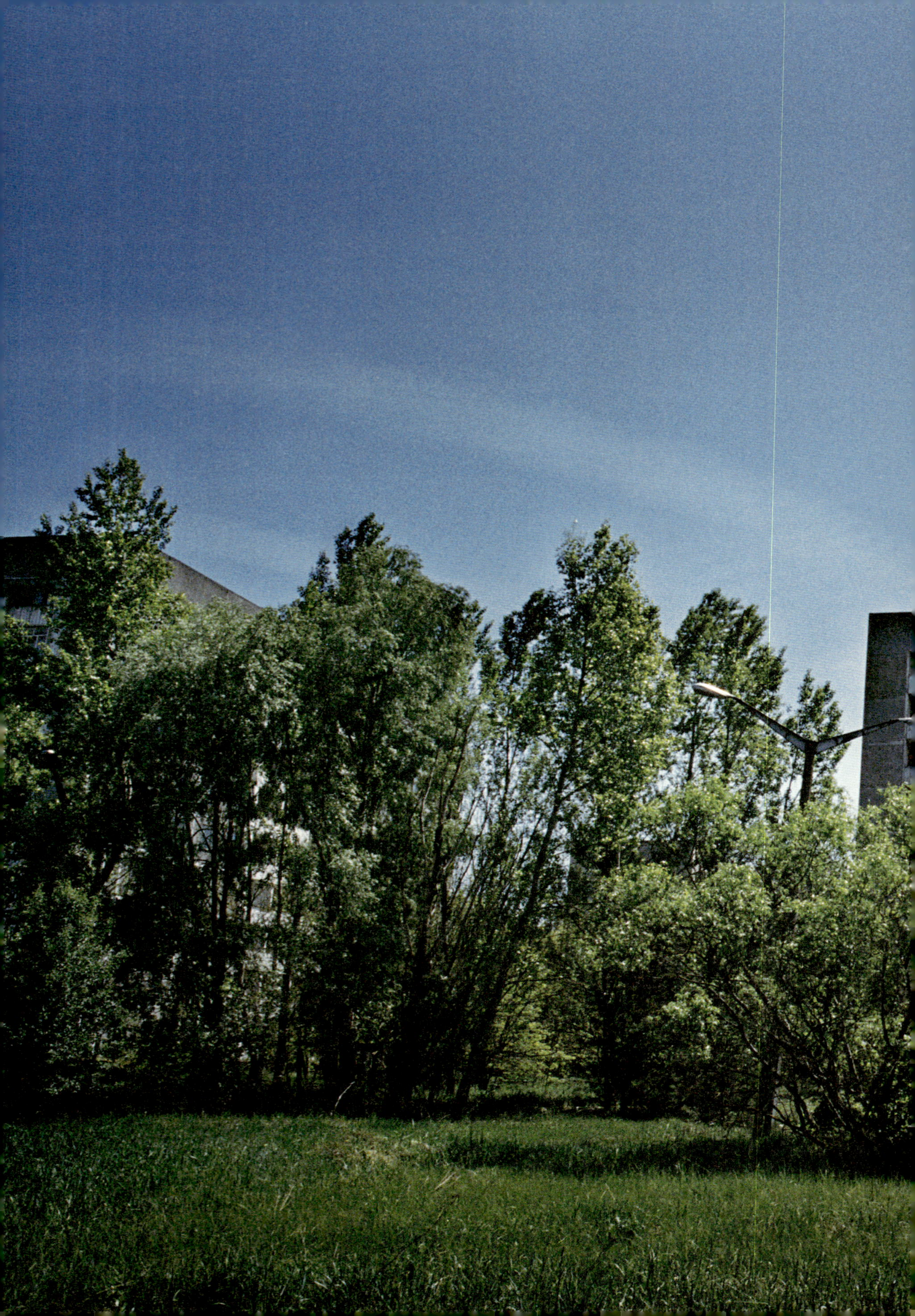

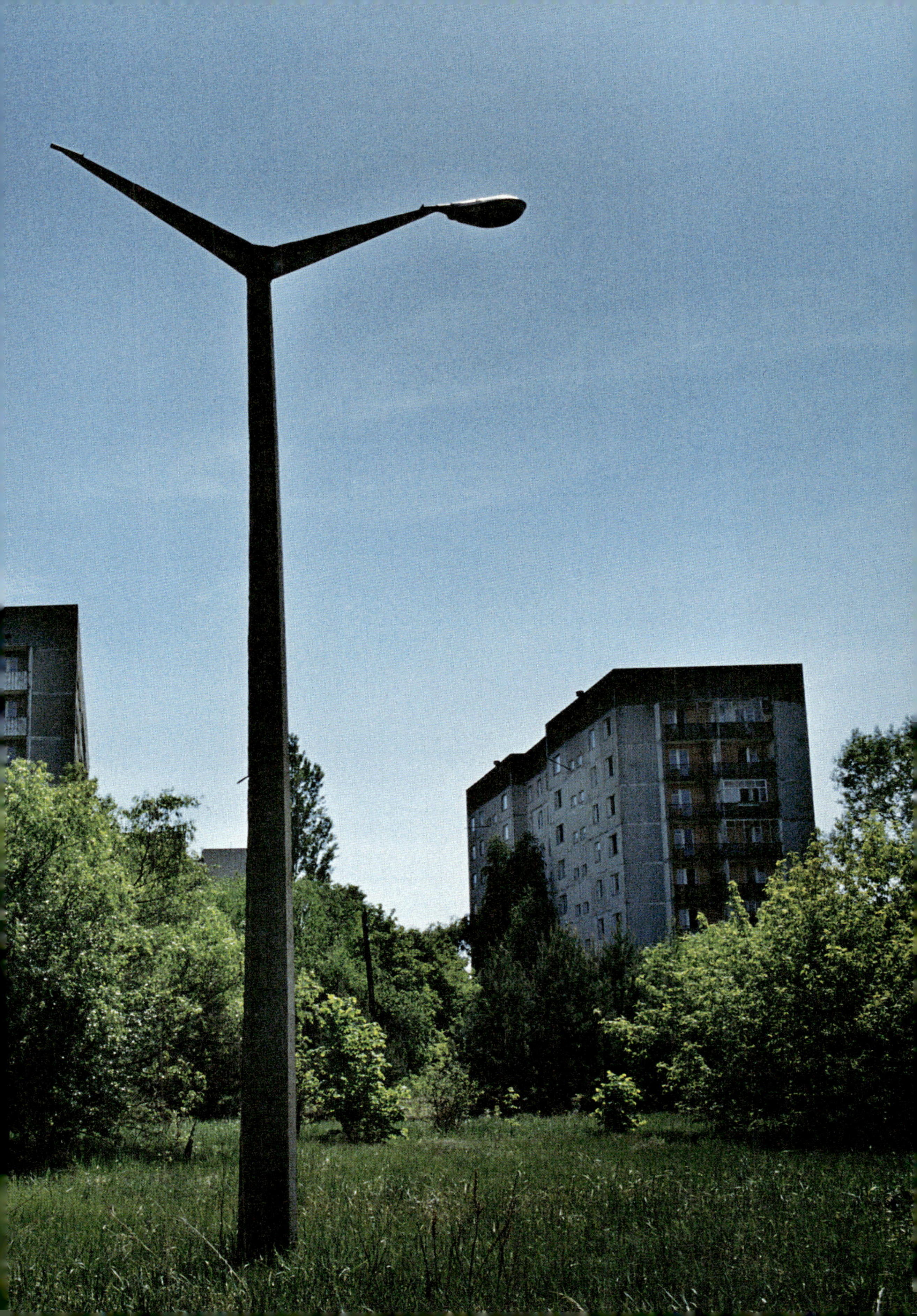

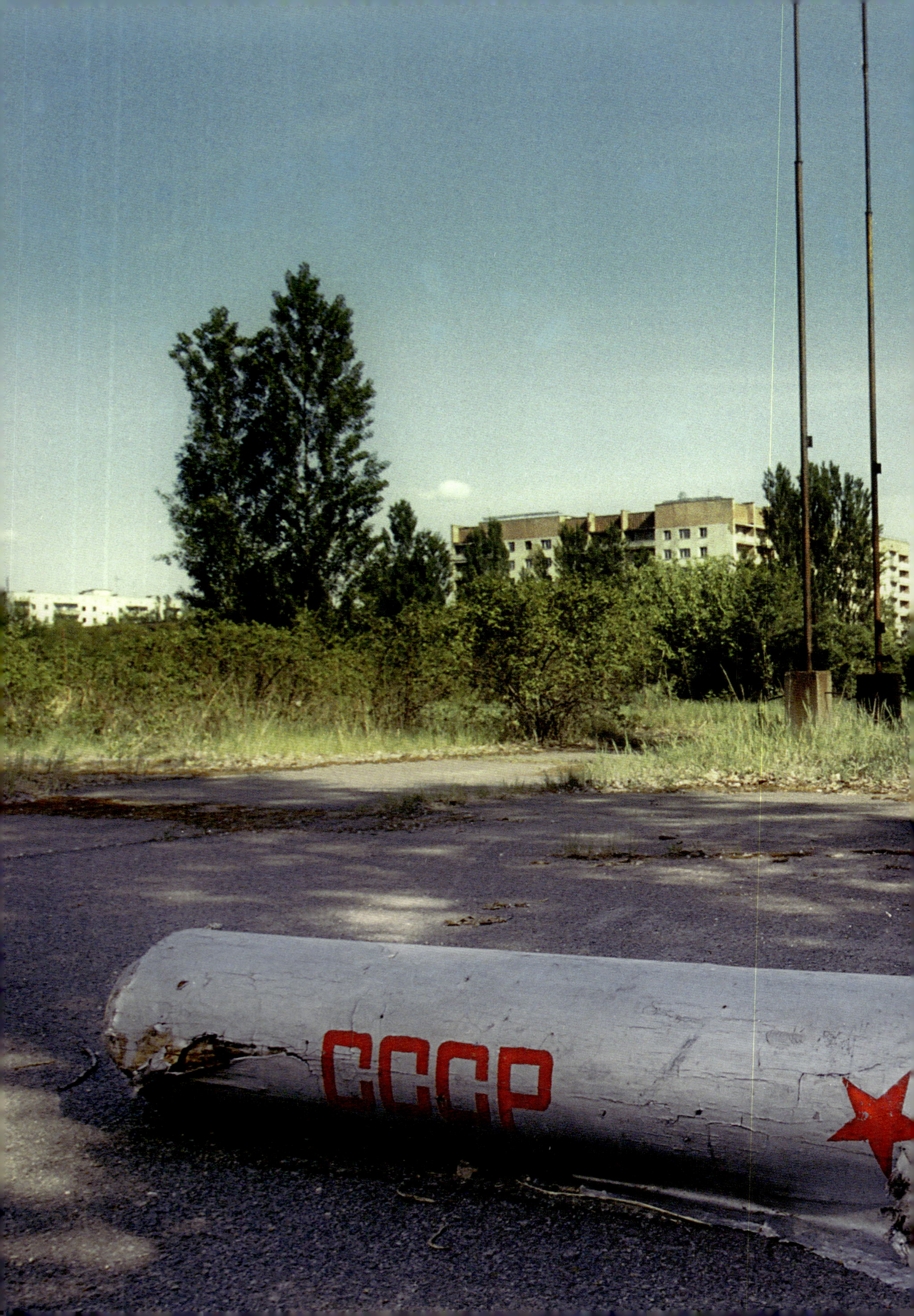
СССР

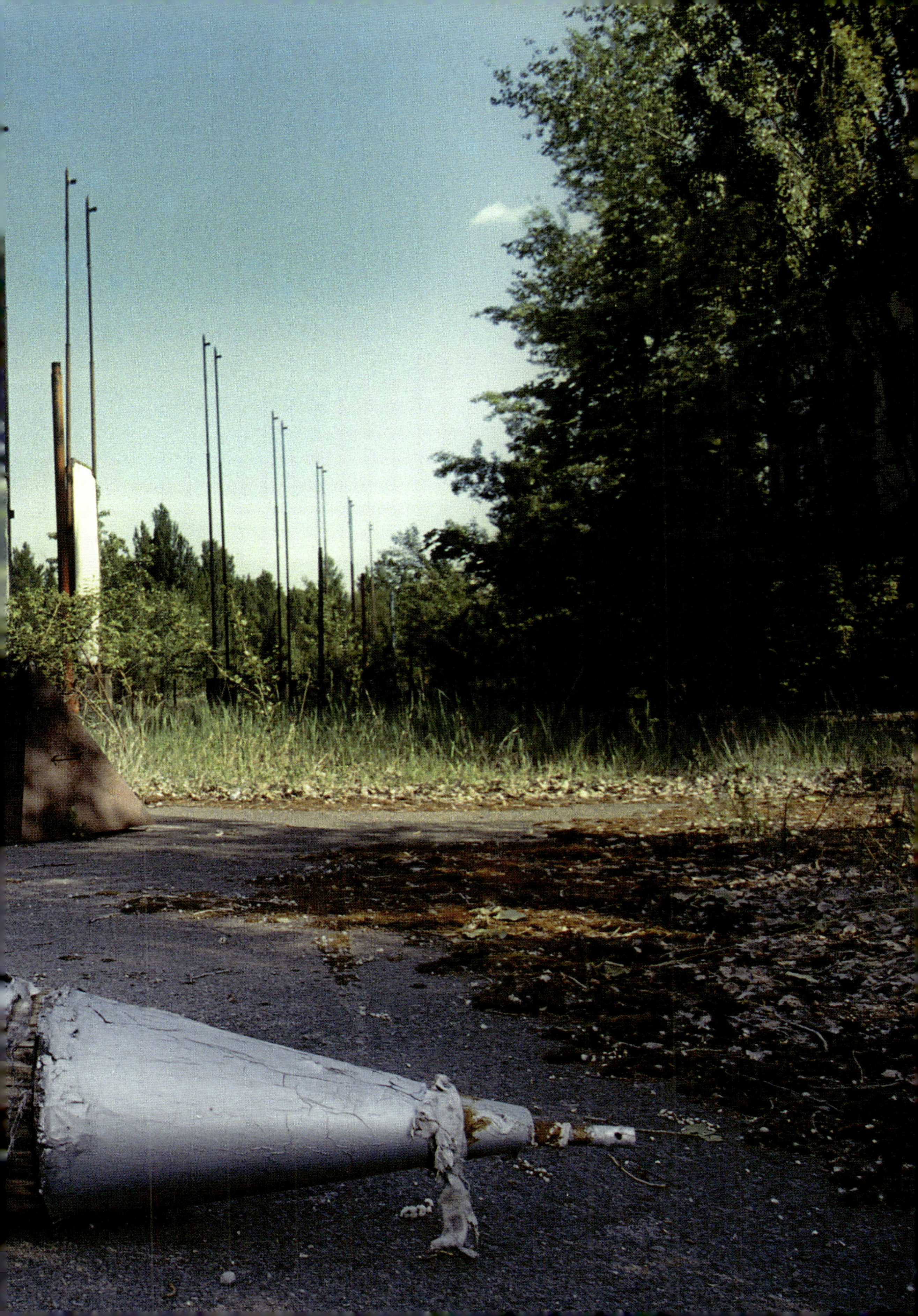

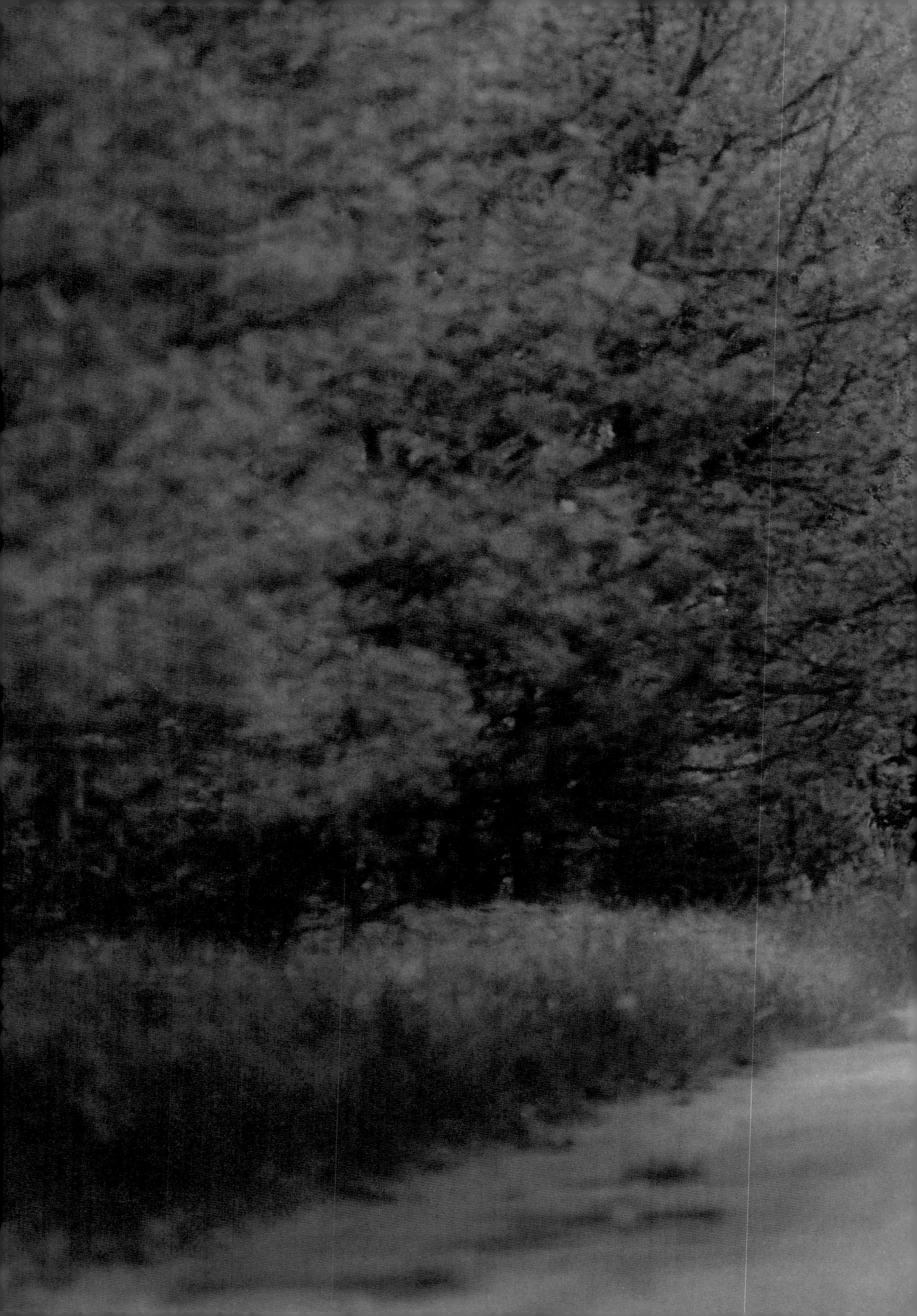

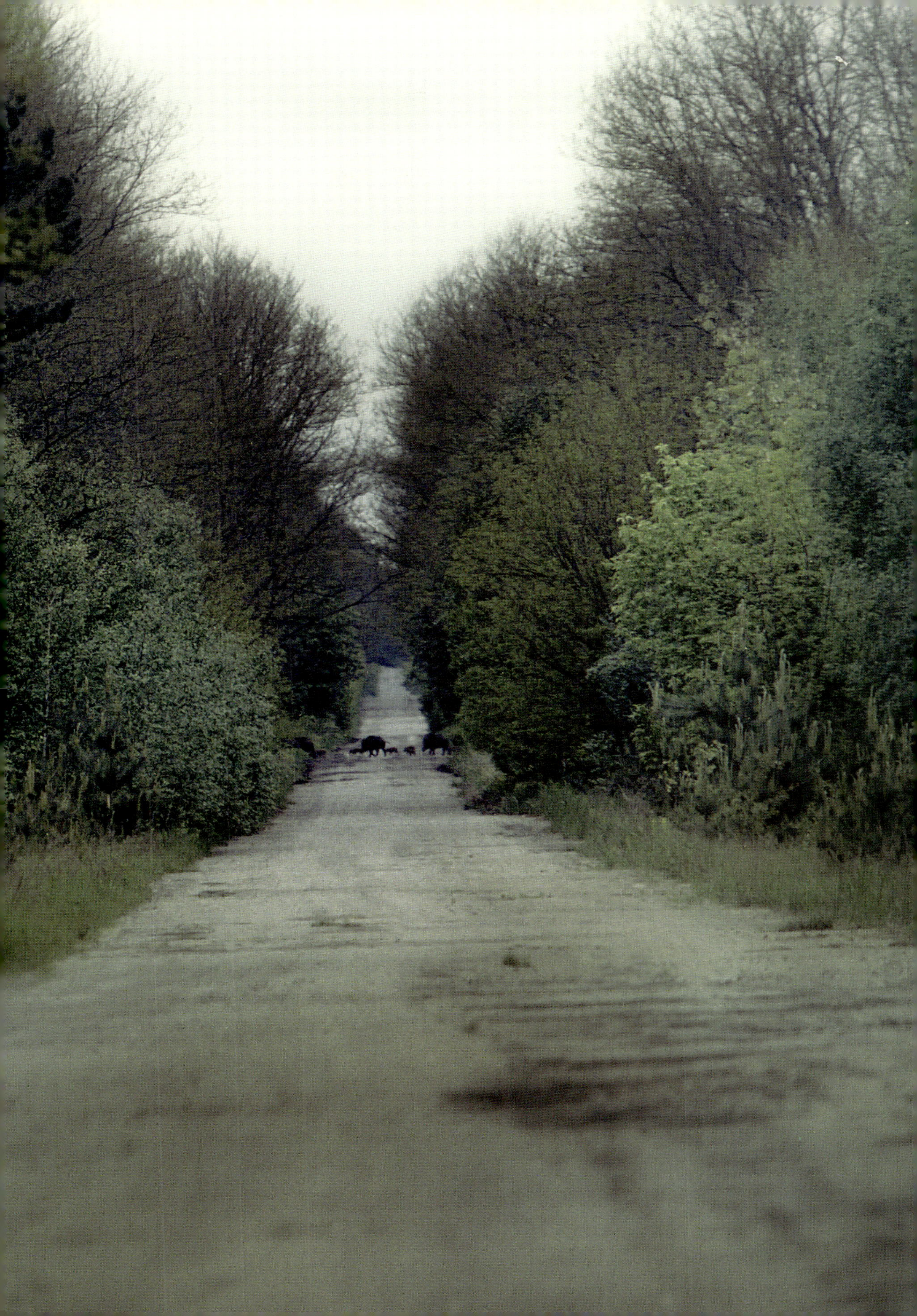

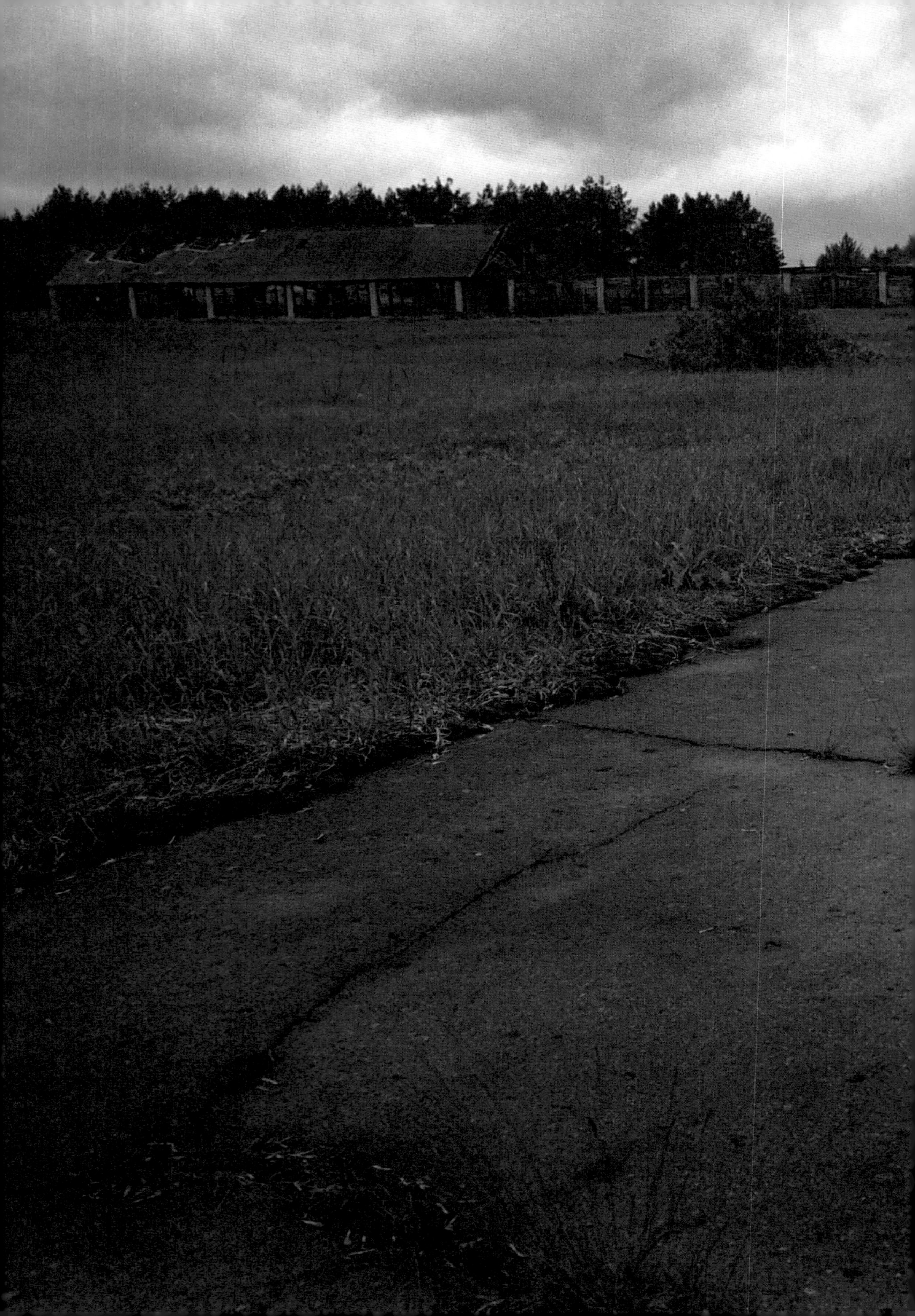

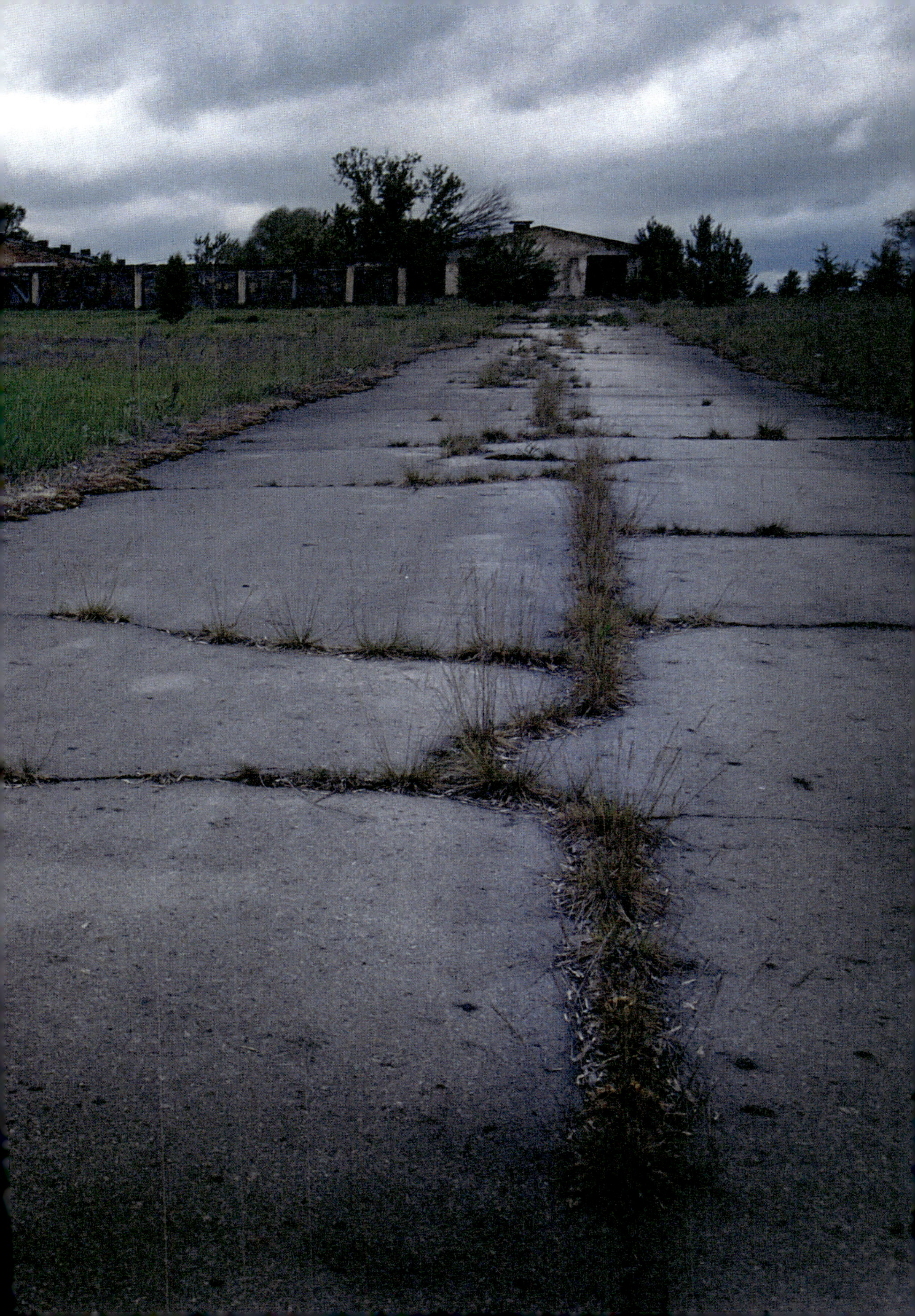

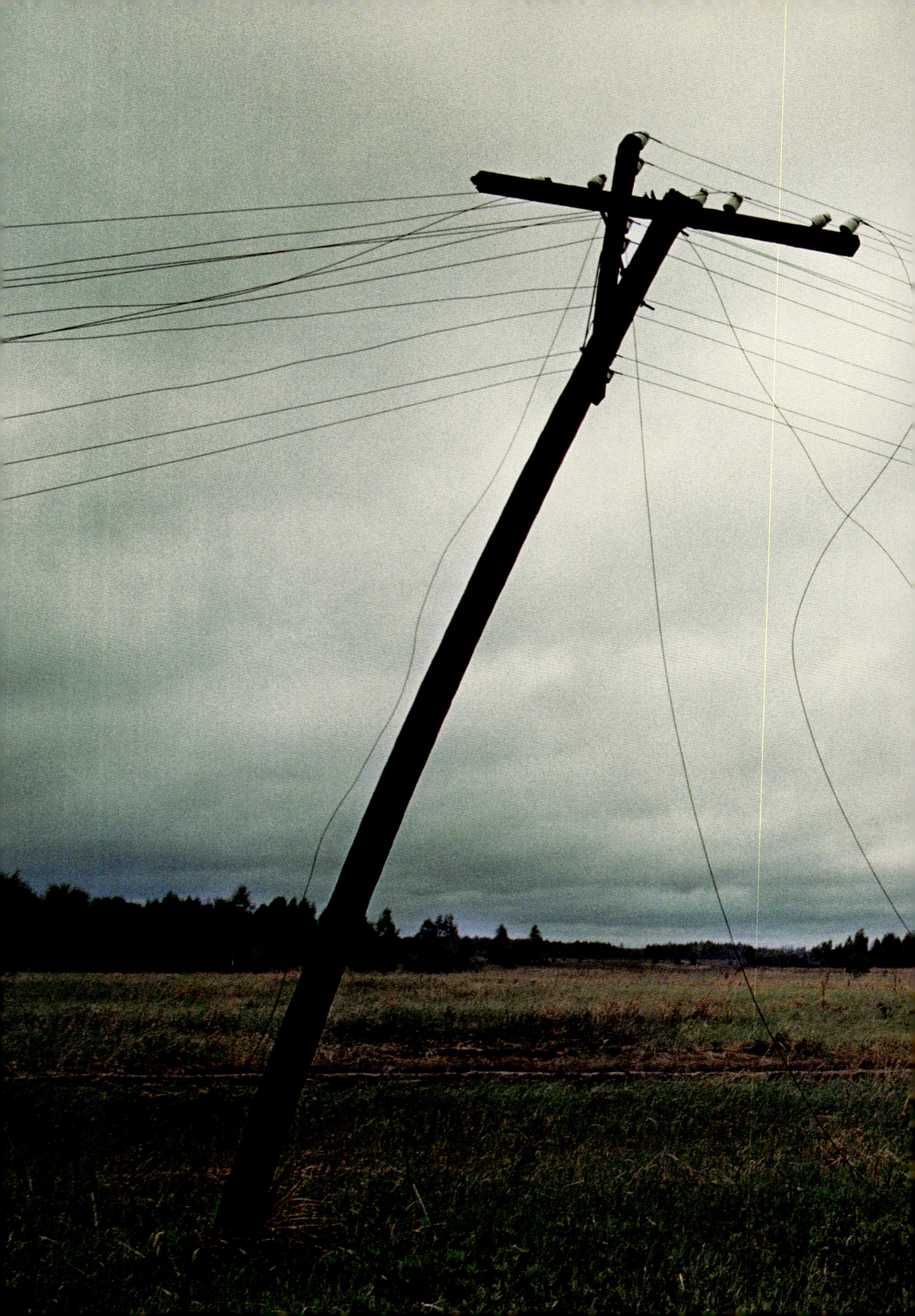

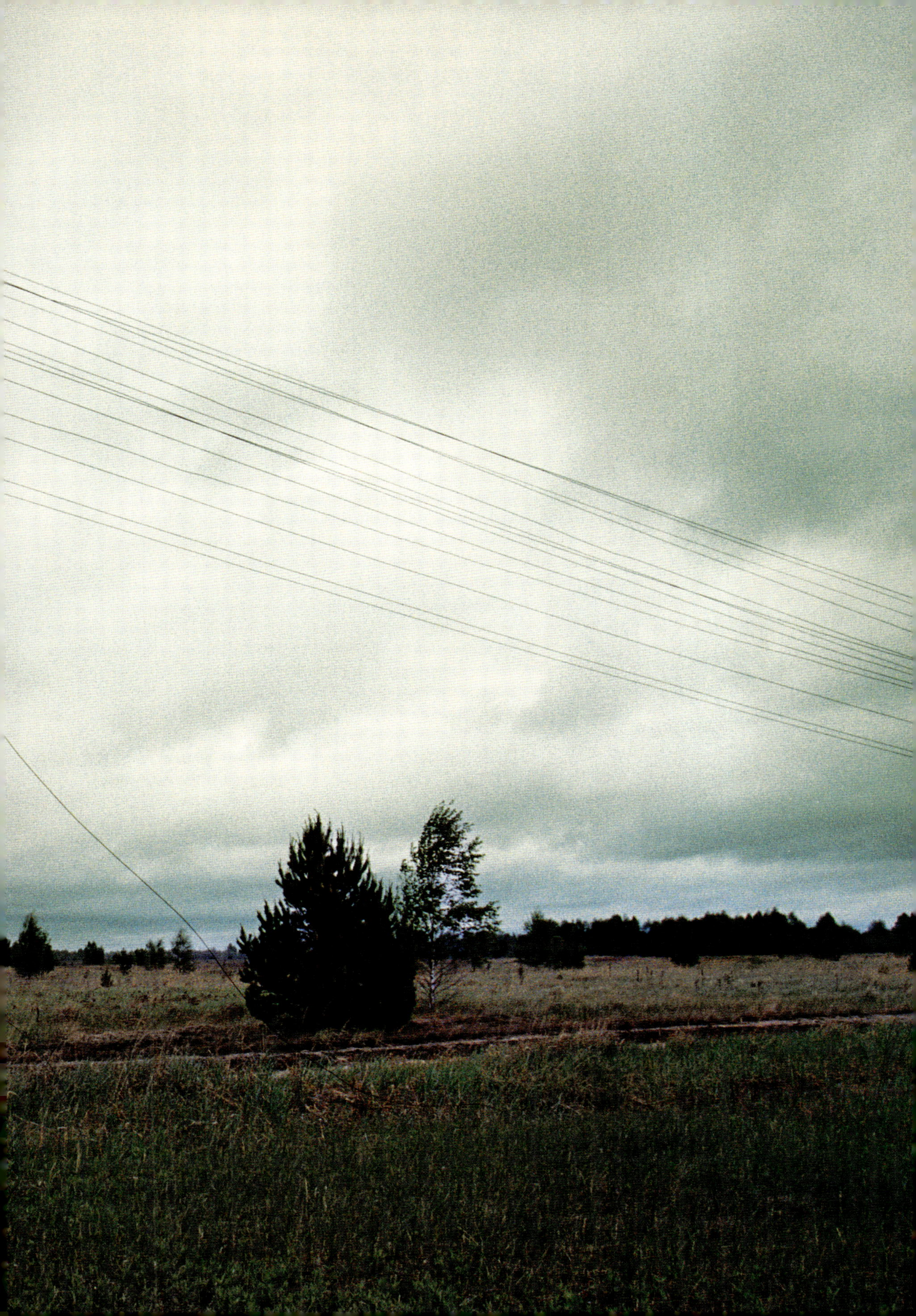

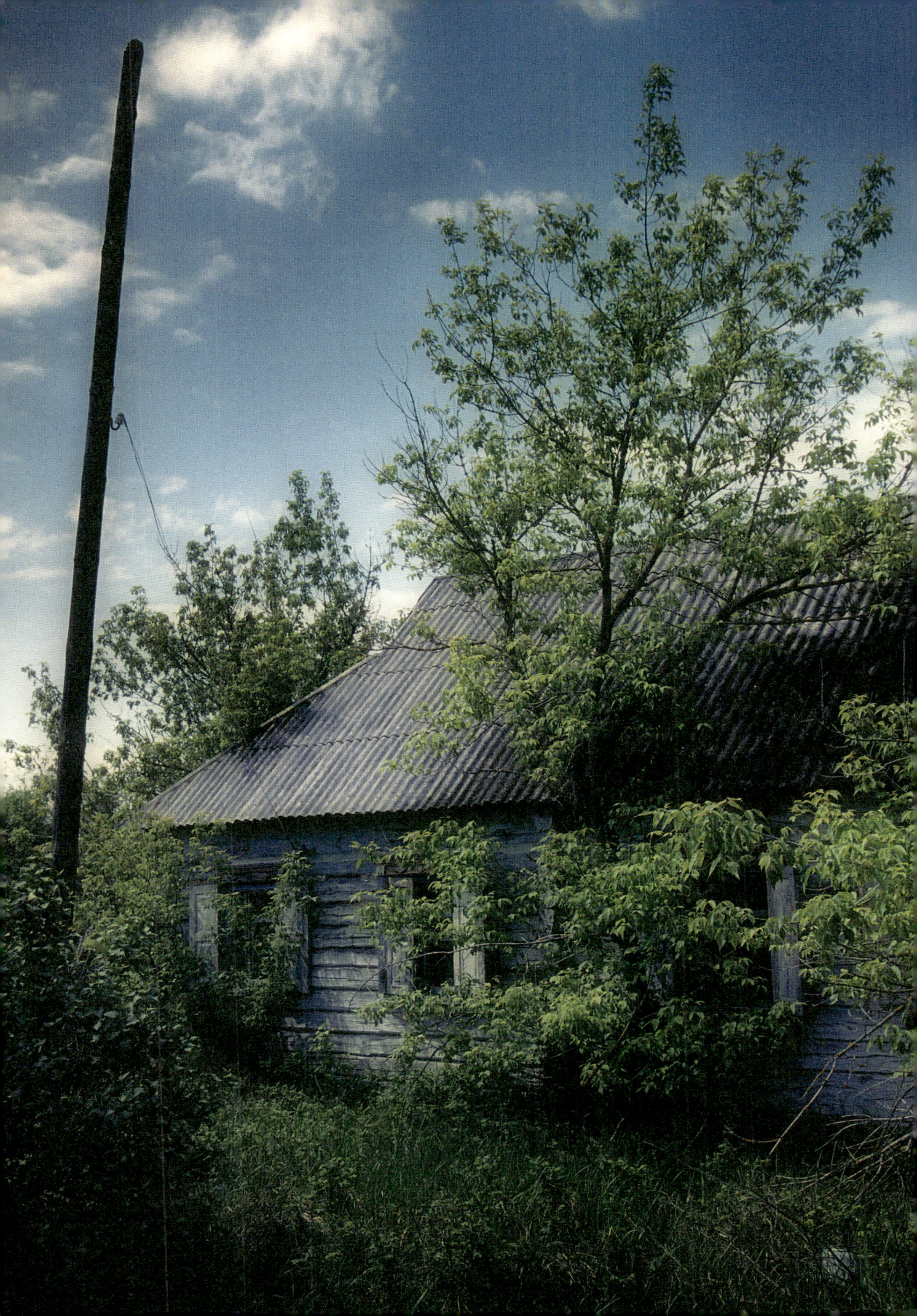

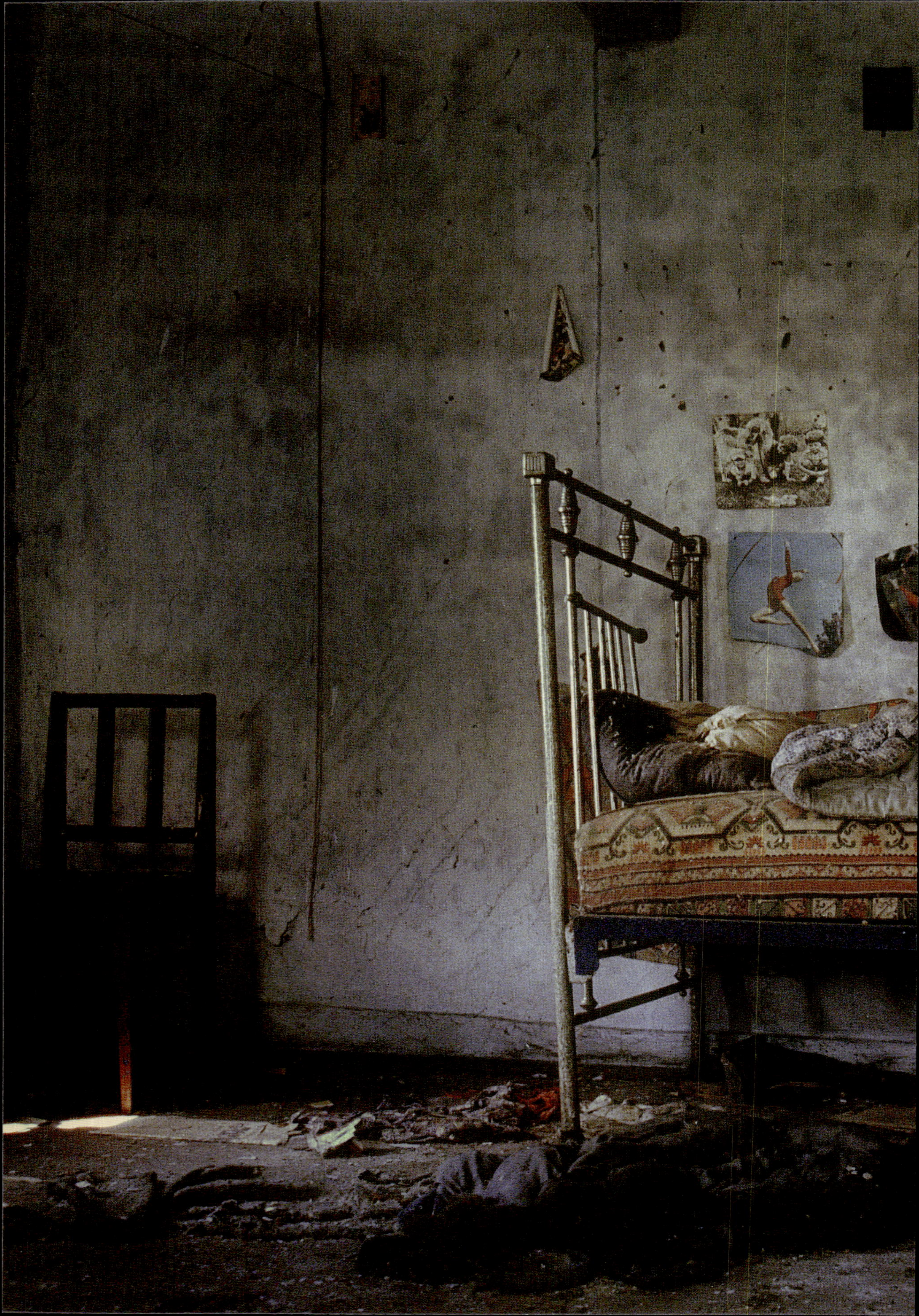

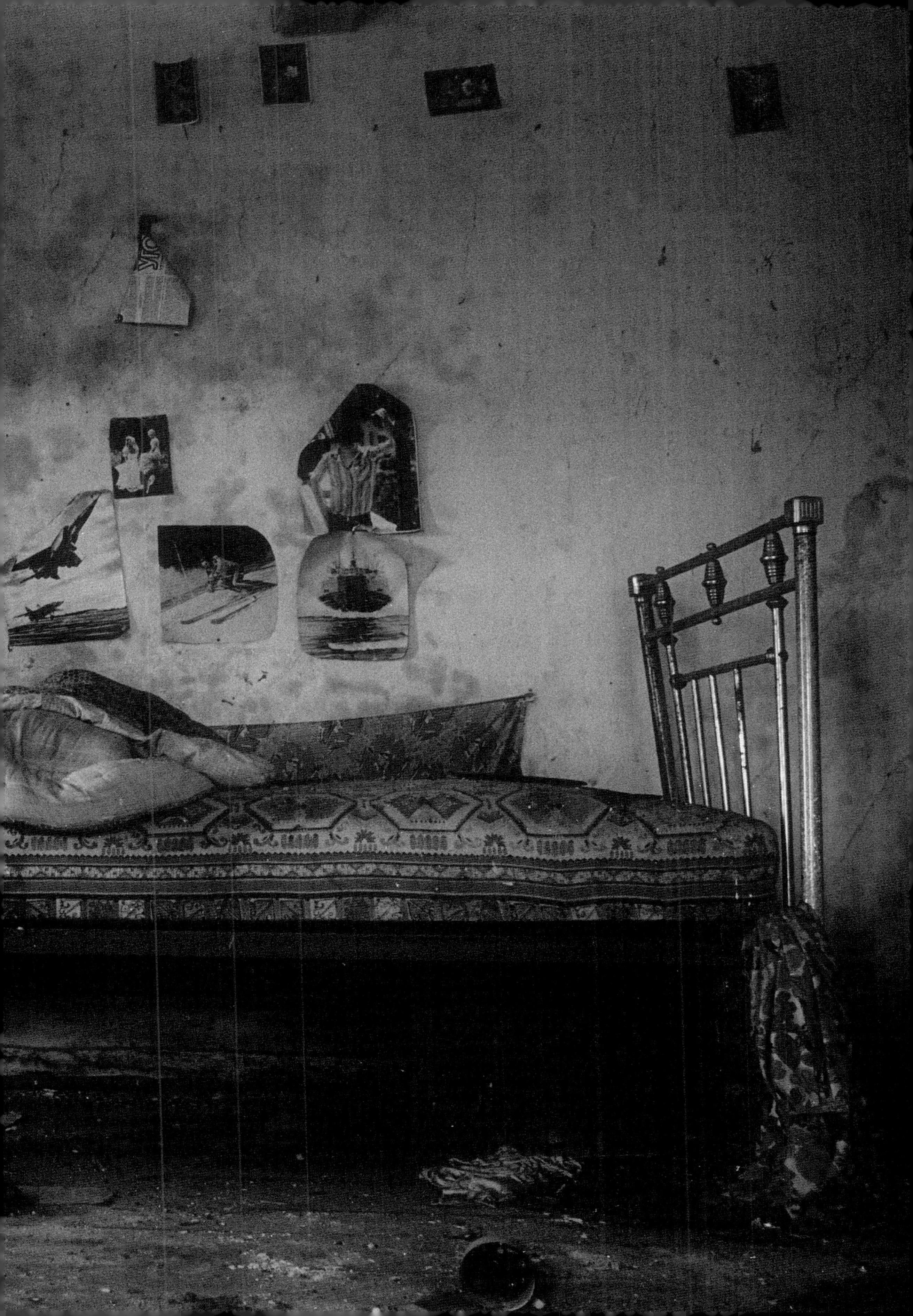

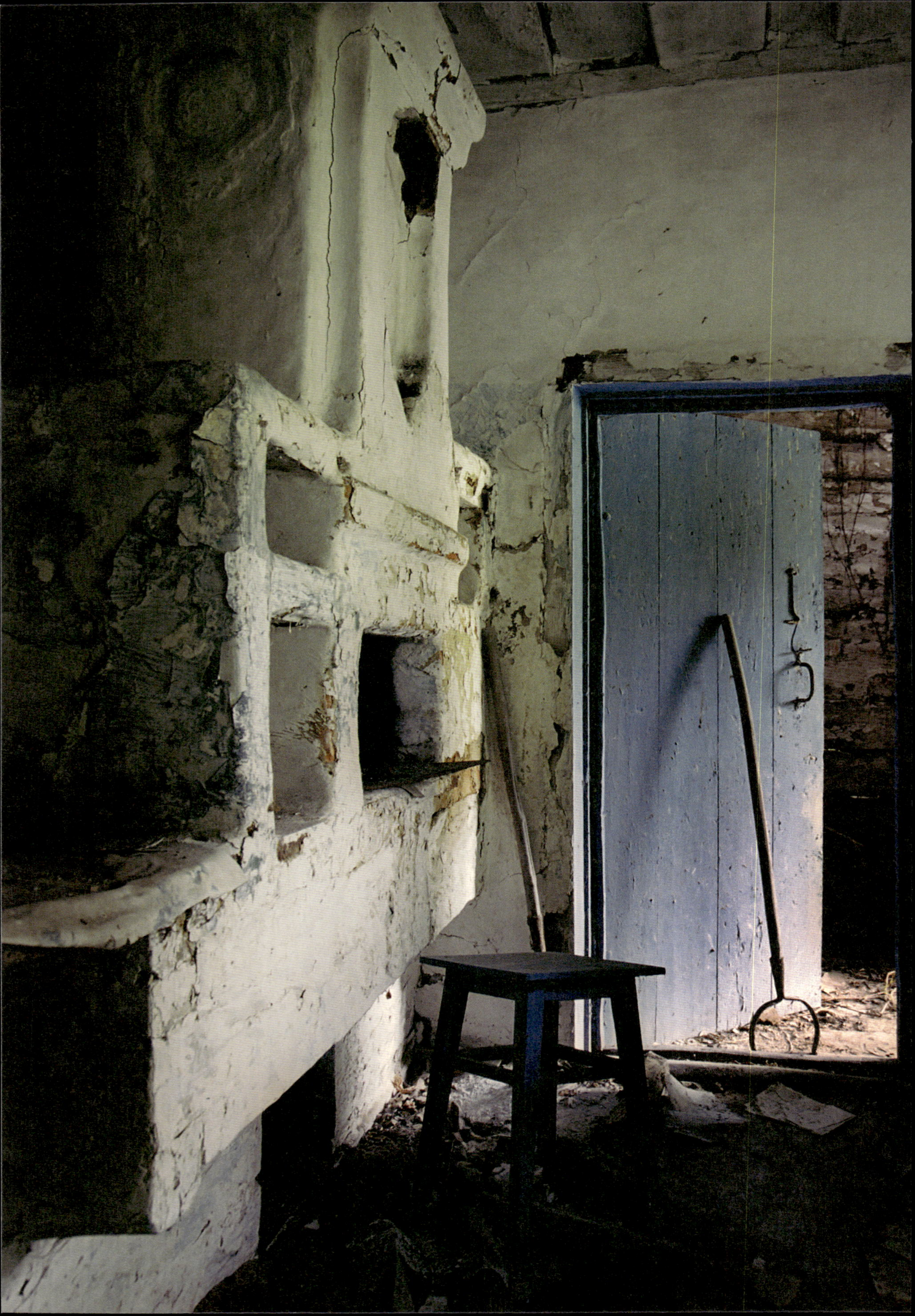

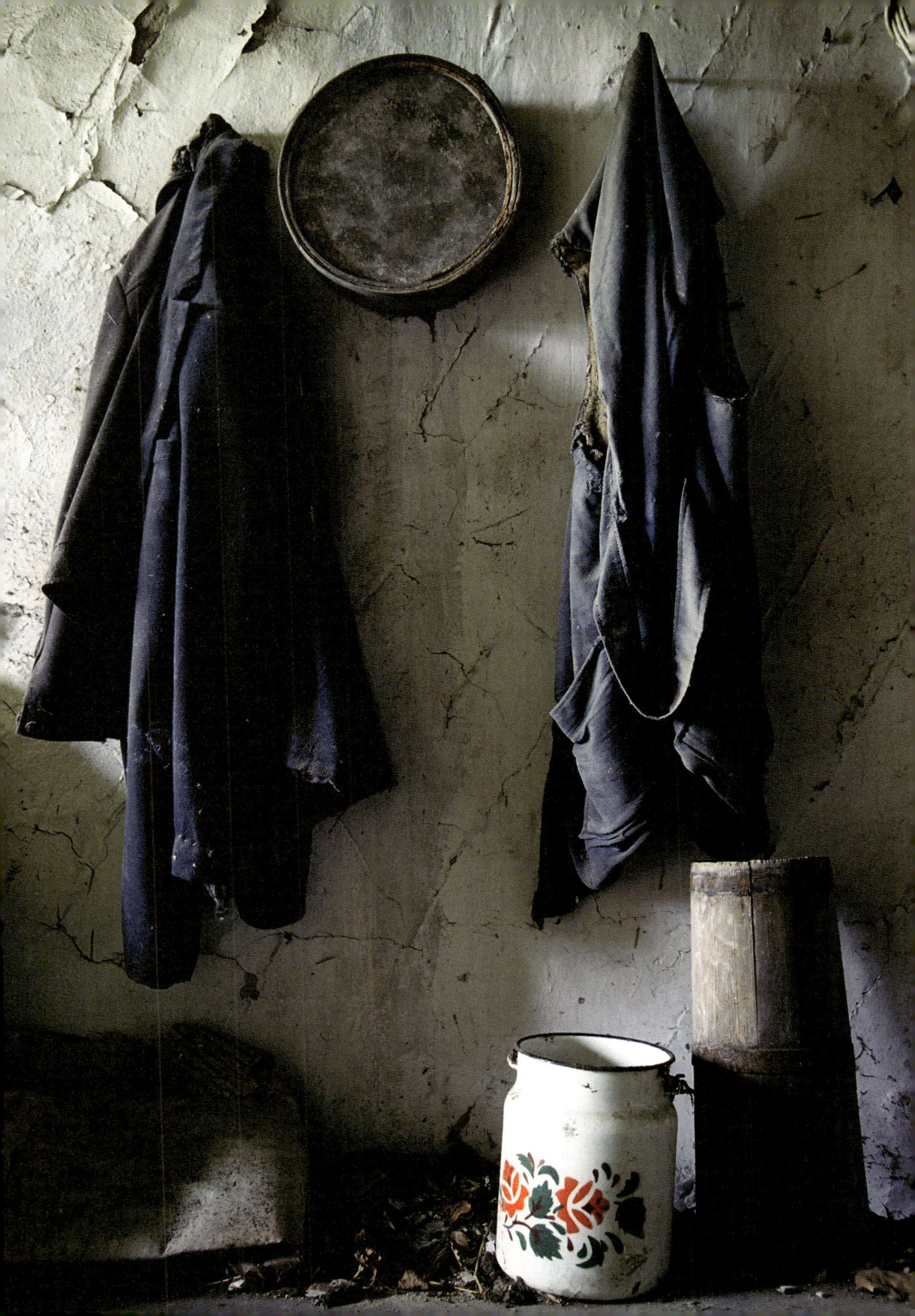

ЛЕНІН
1870
РОКІ

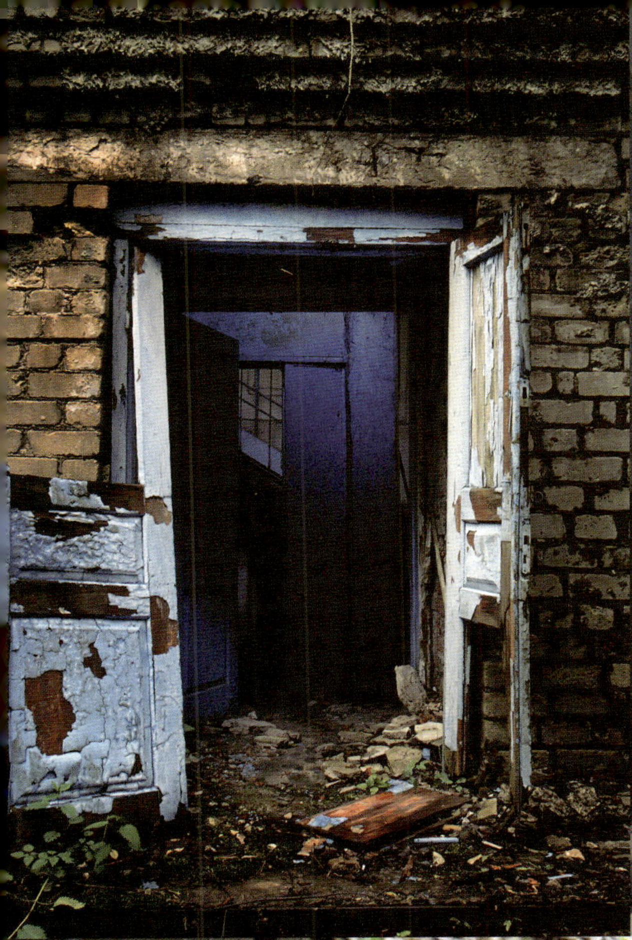
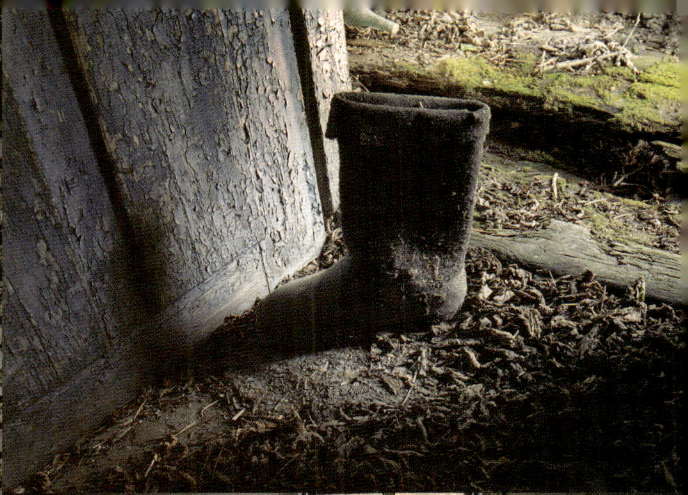
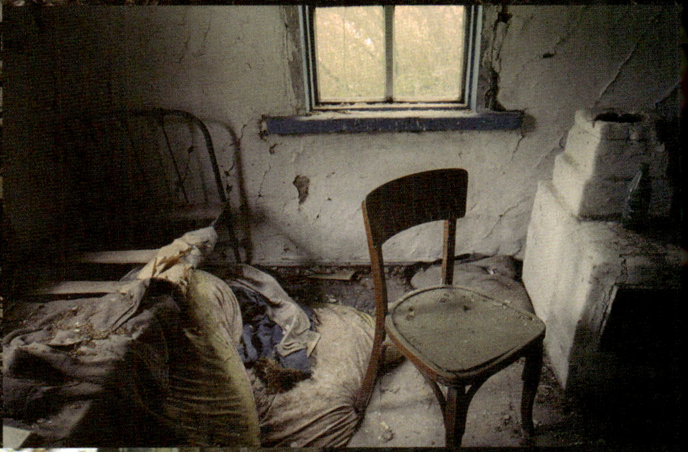
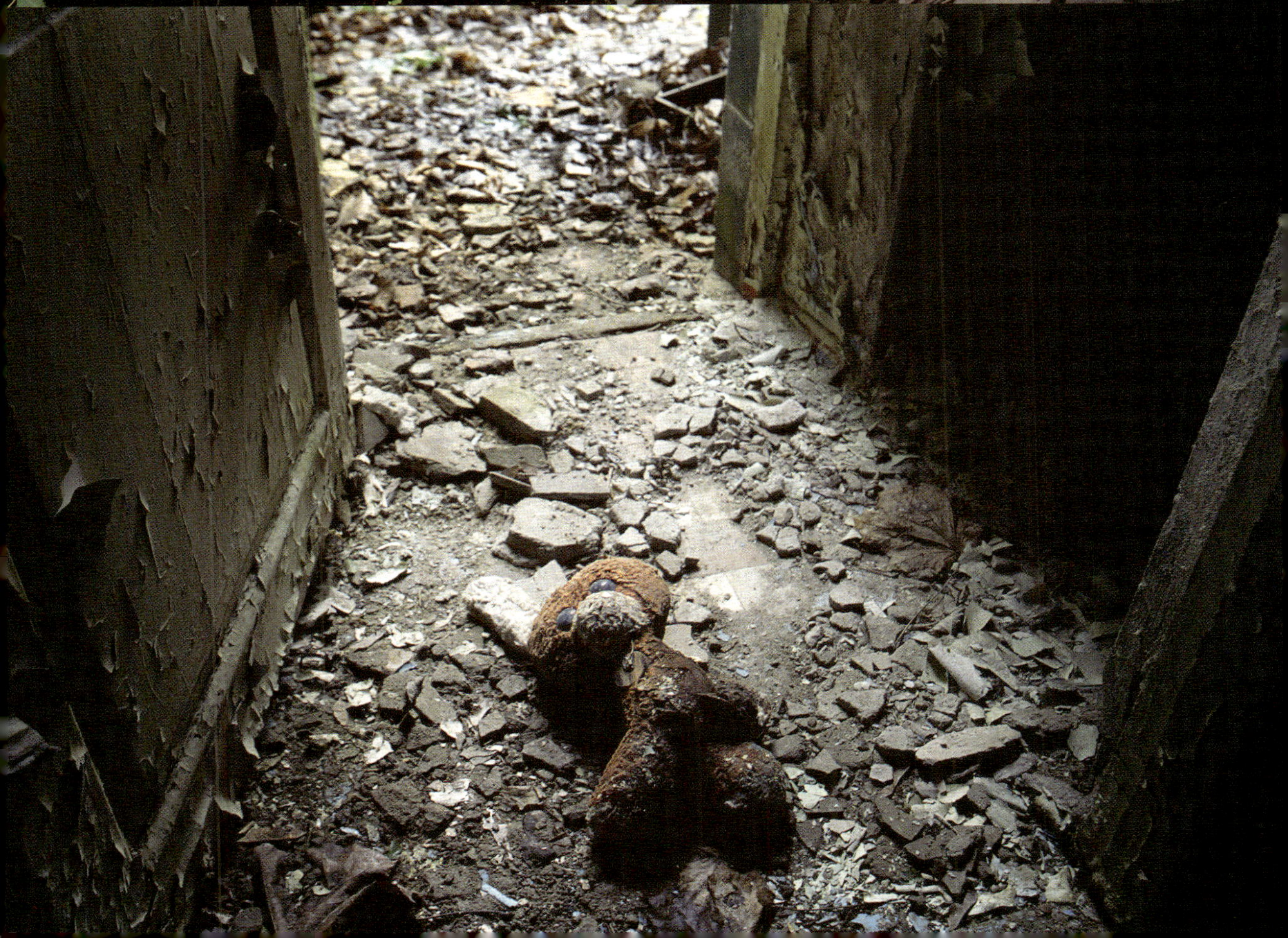

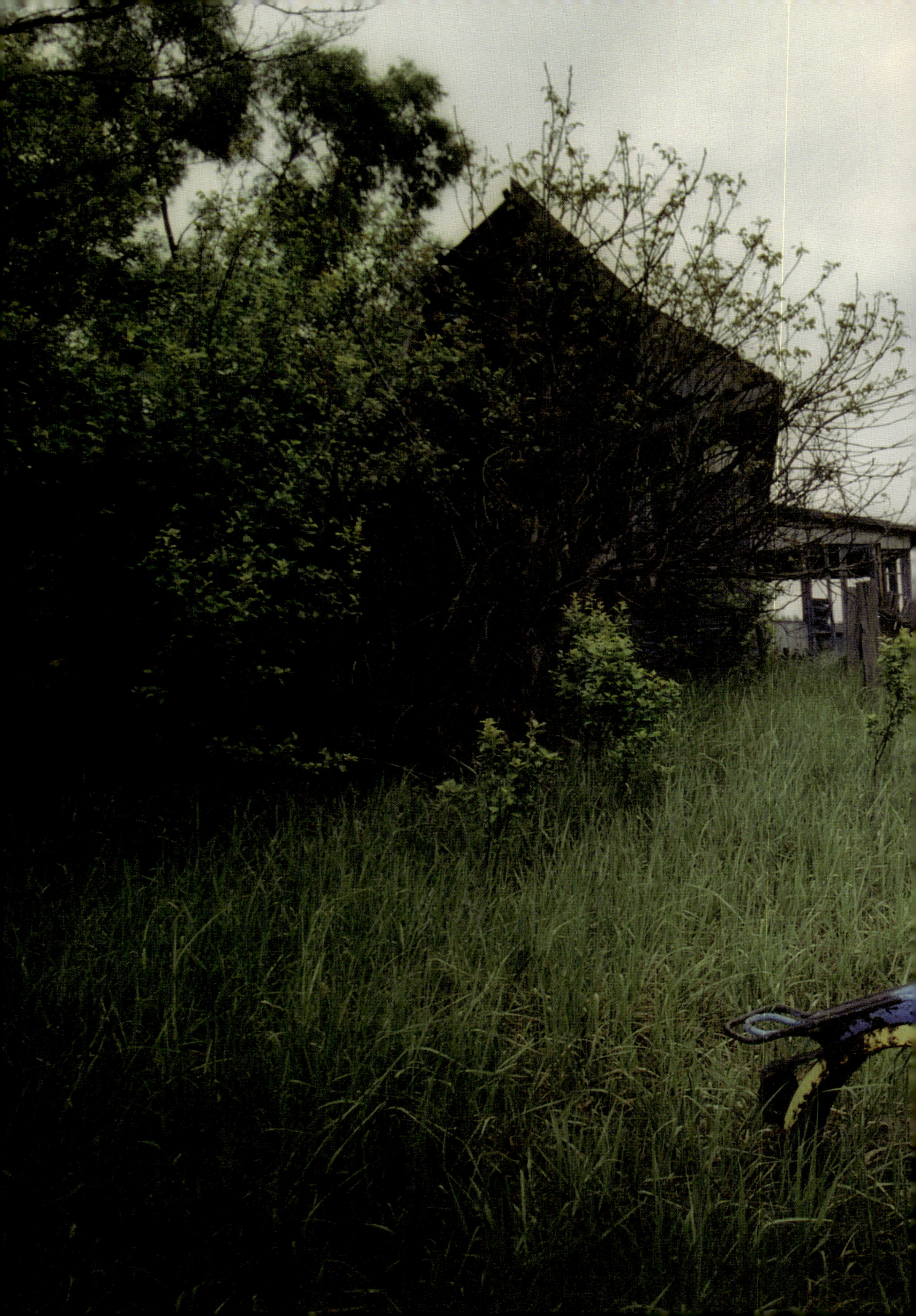

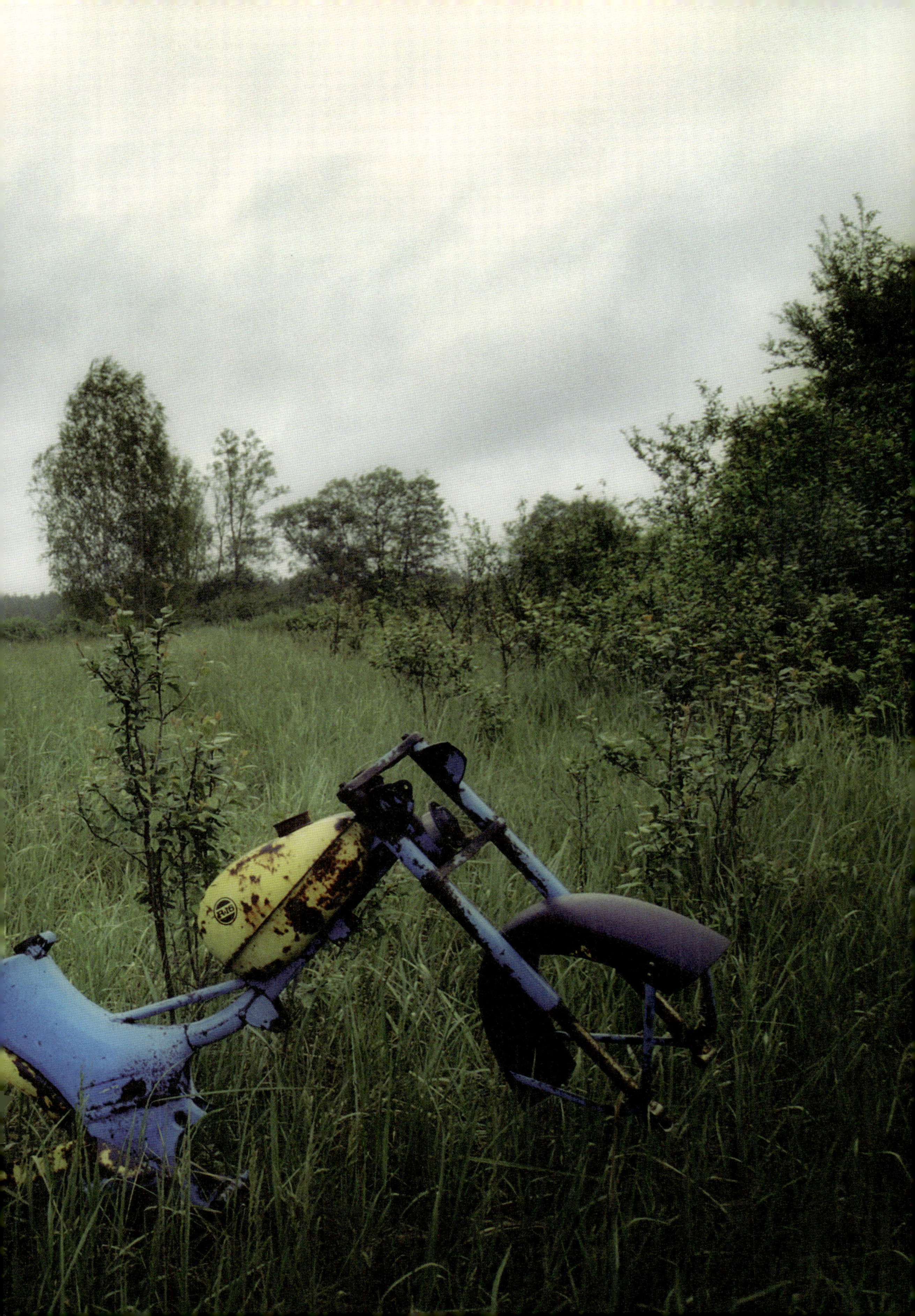

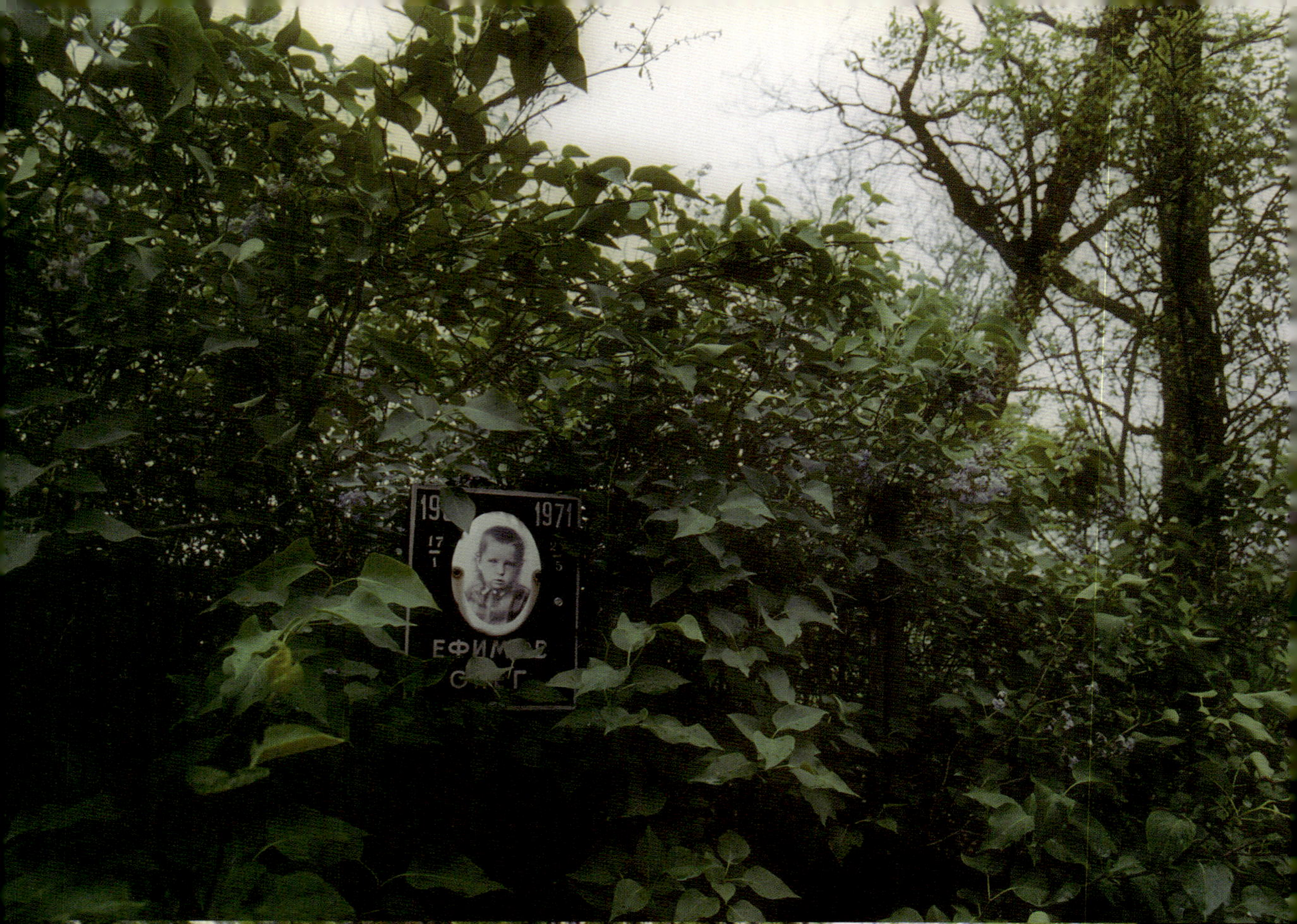
1971
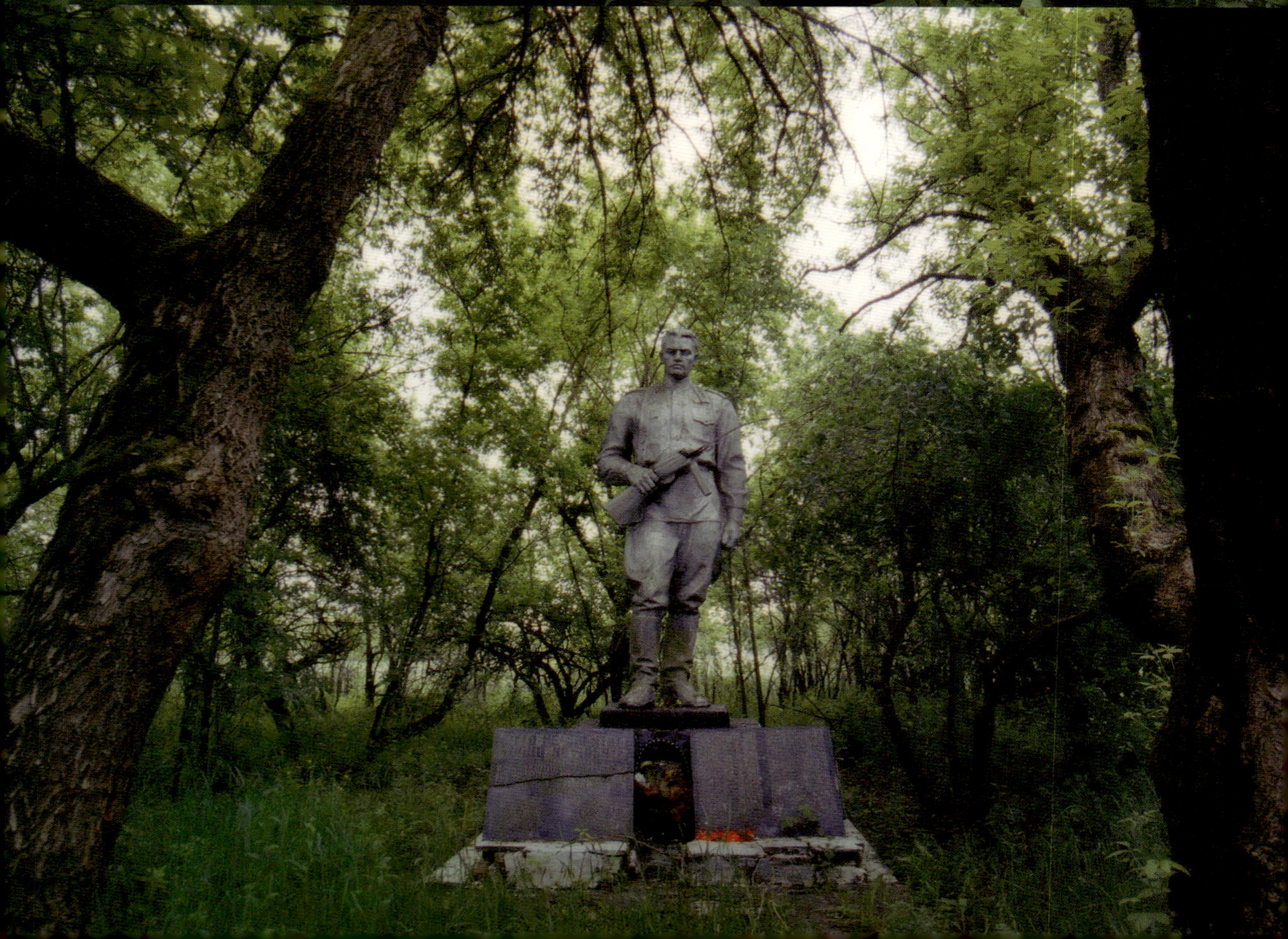

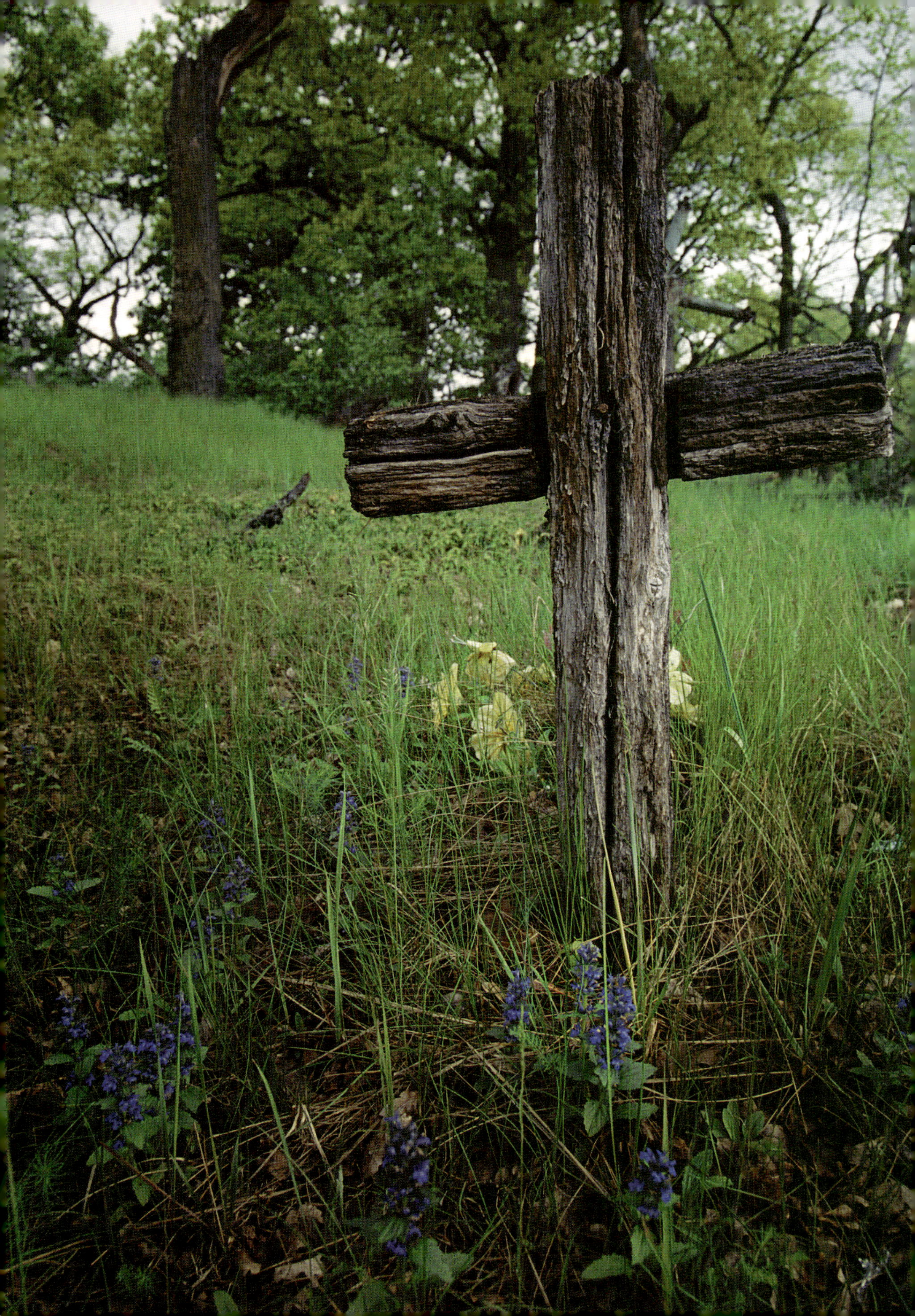

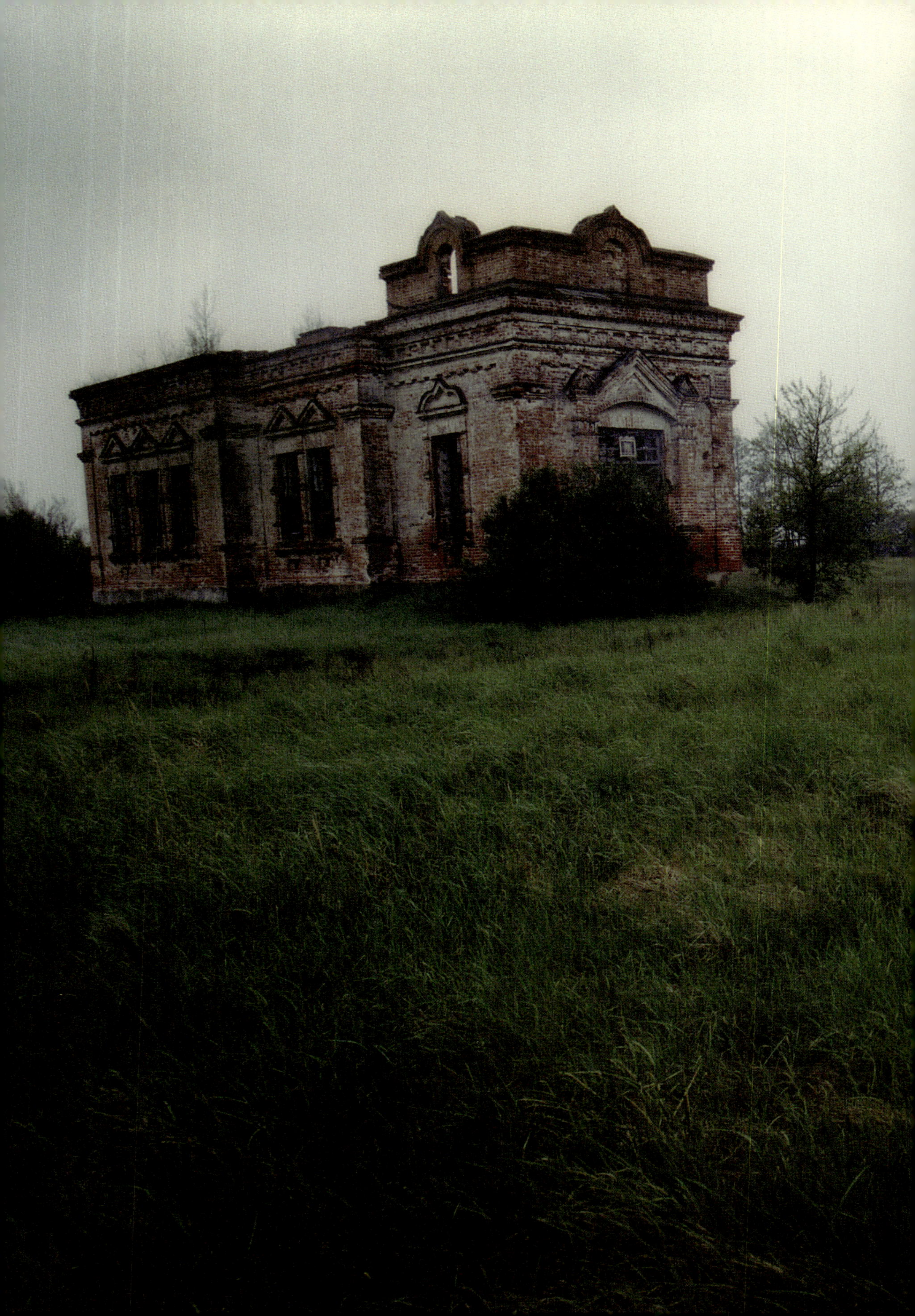

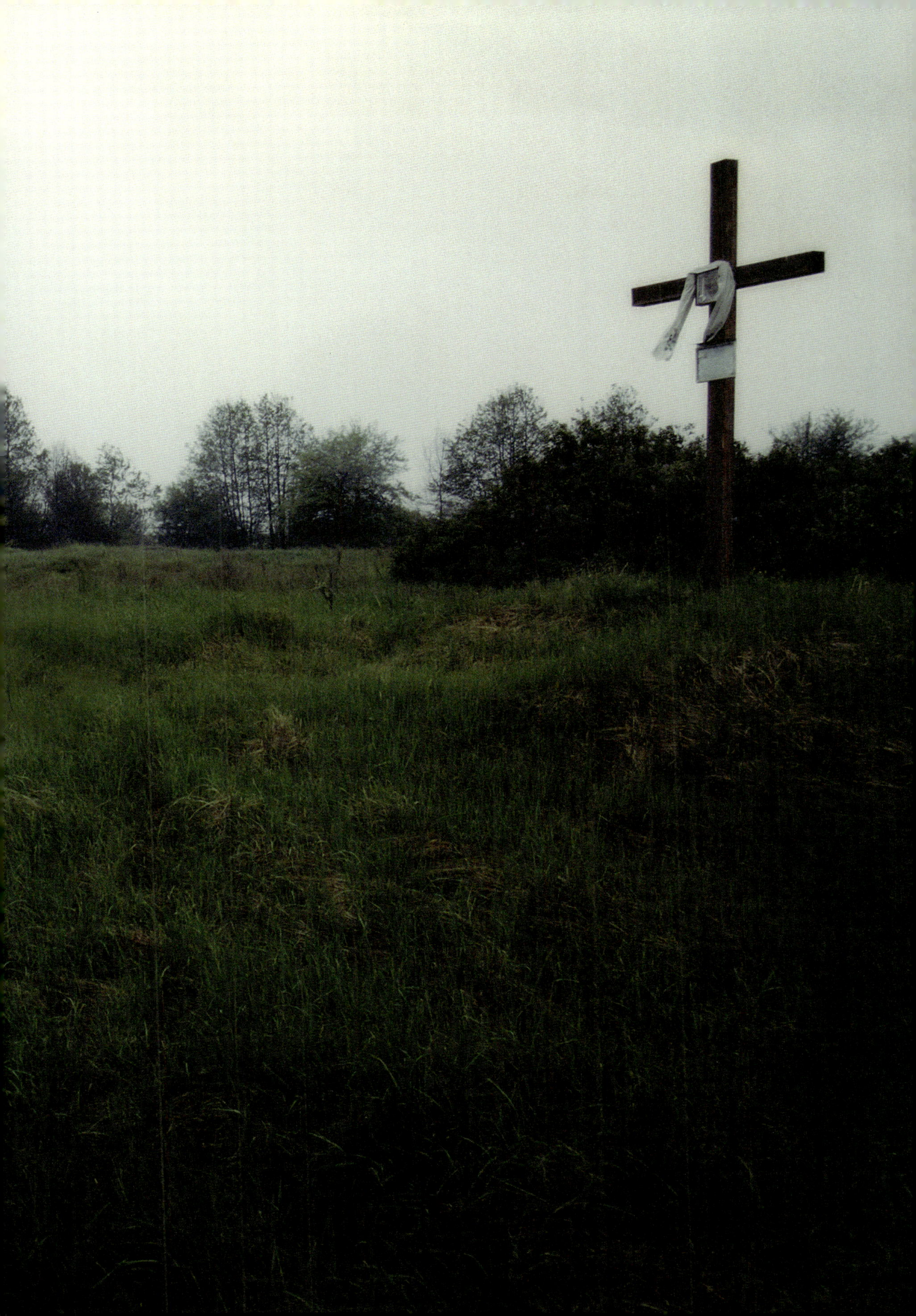

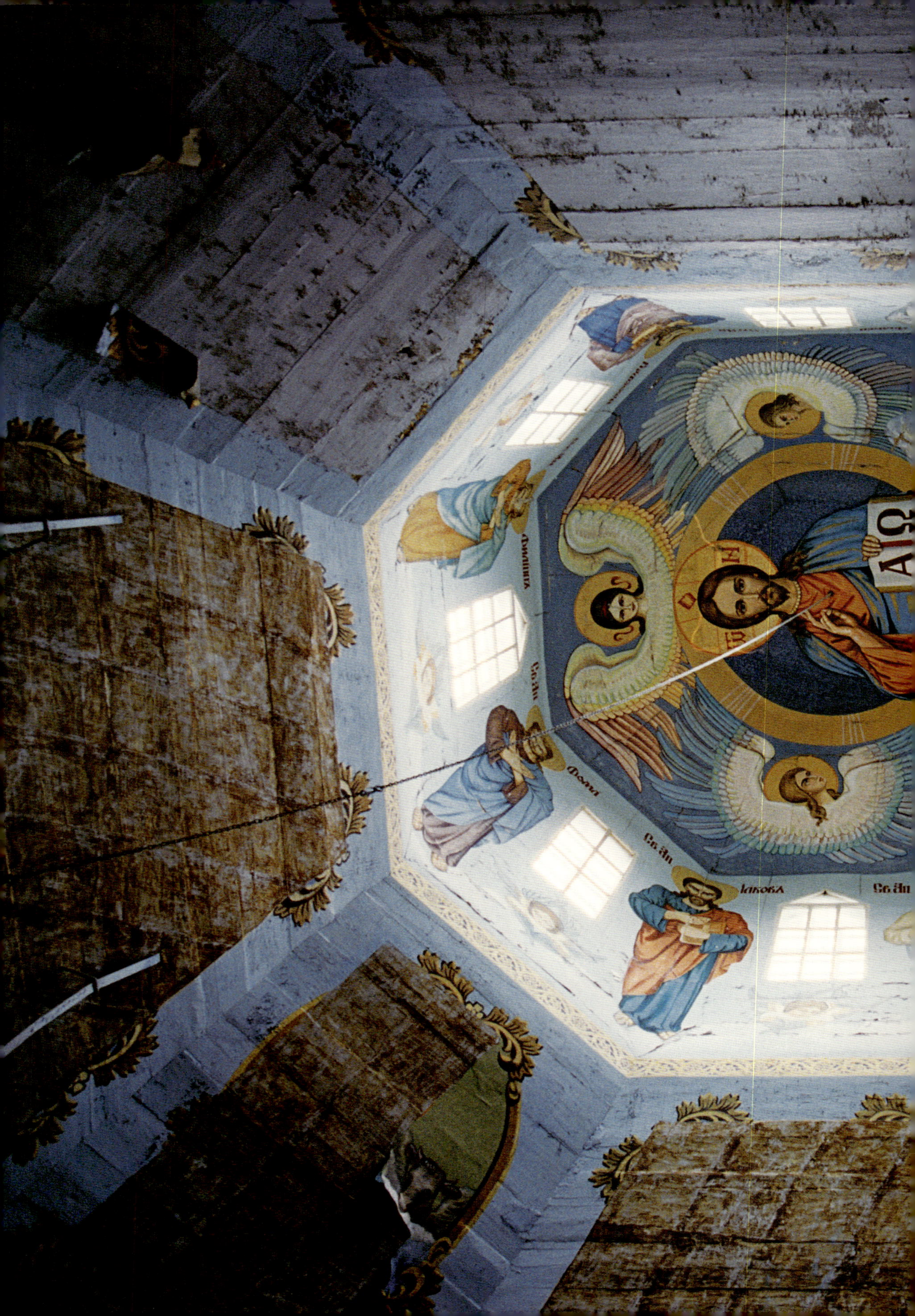
А Ω

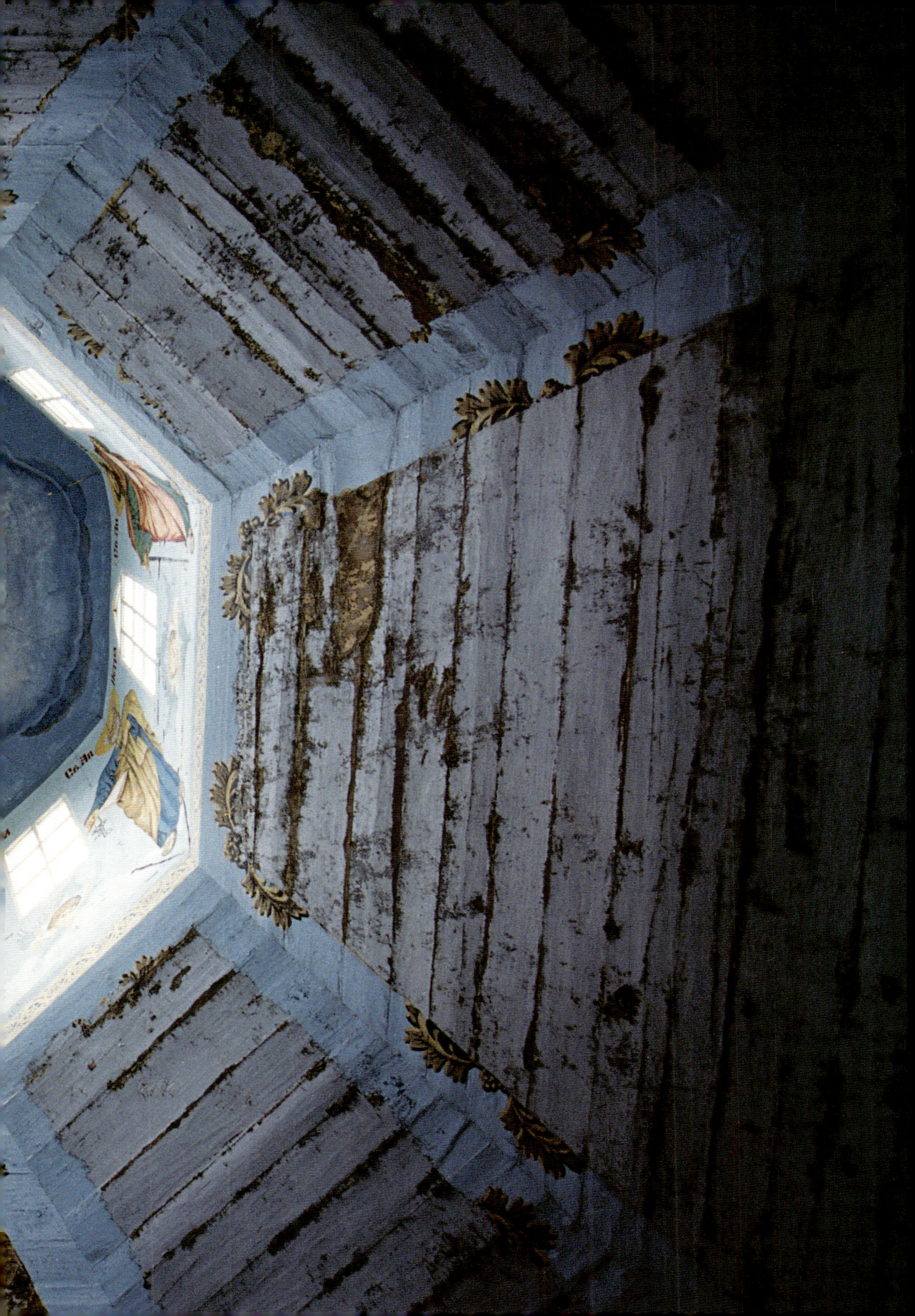

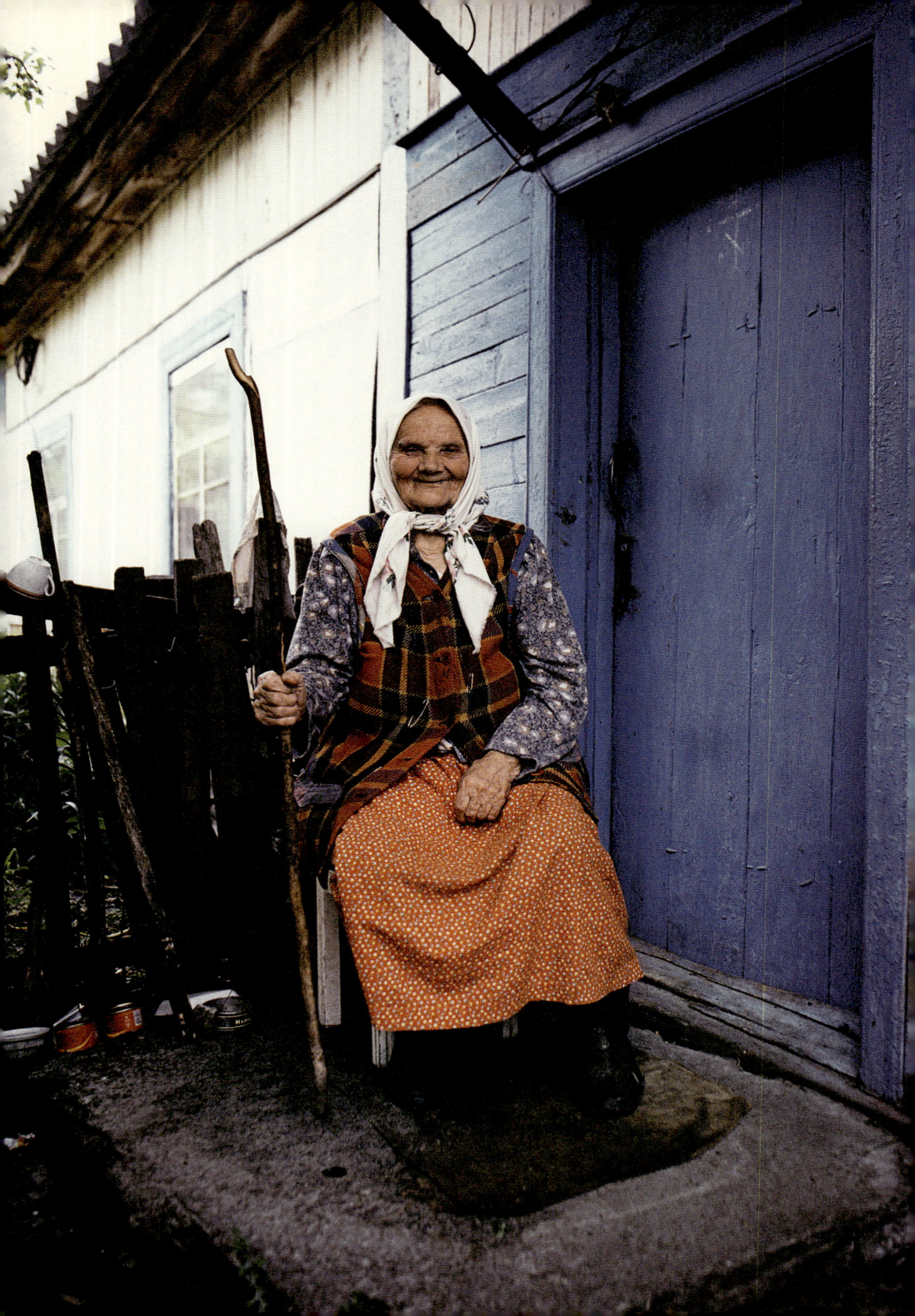

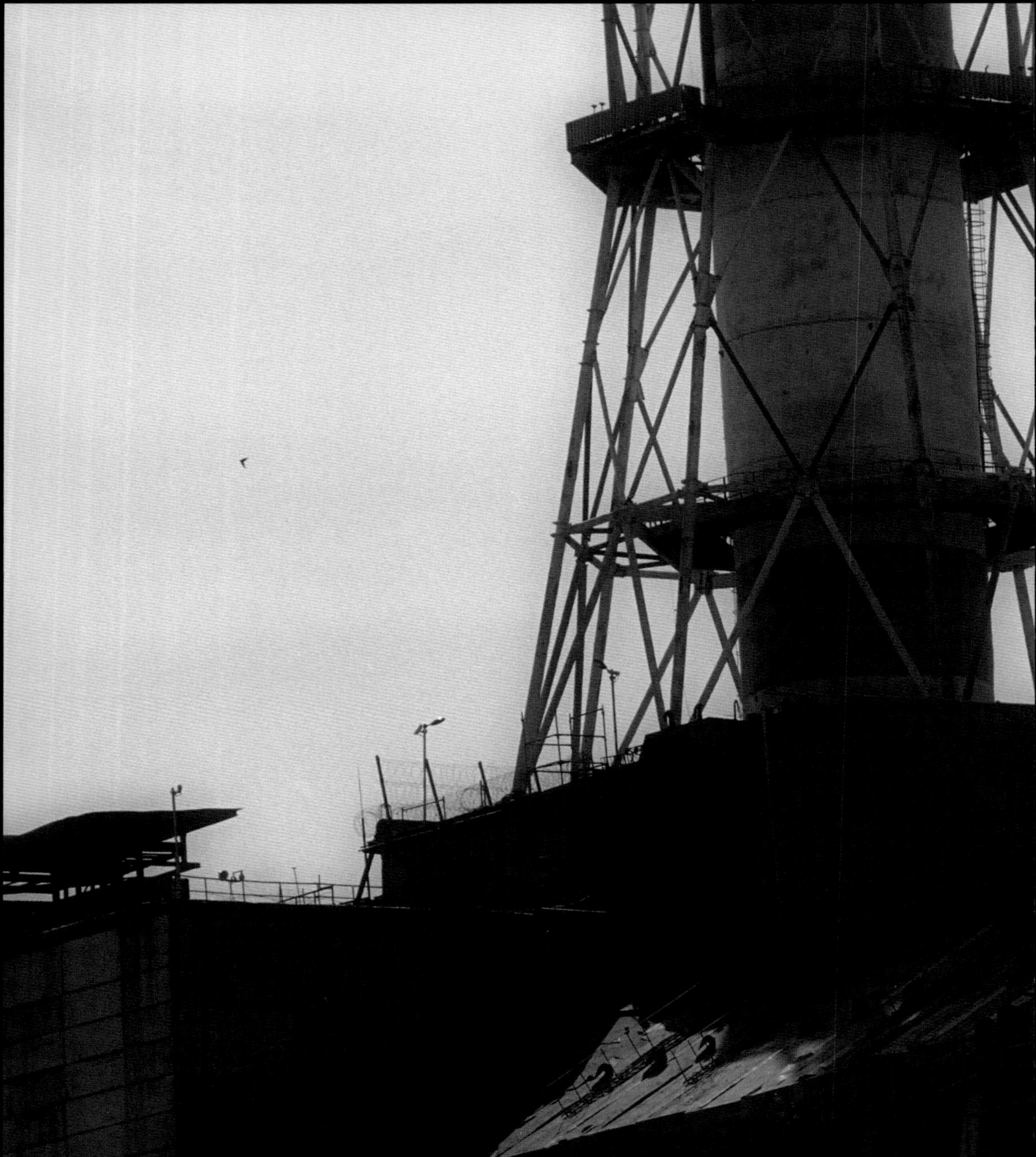

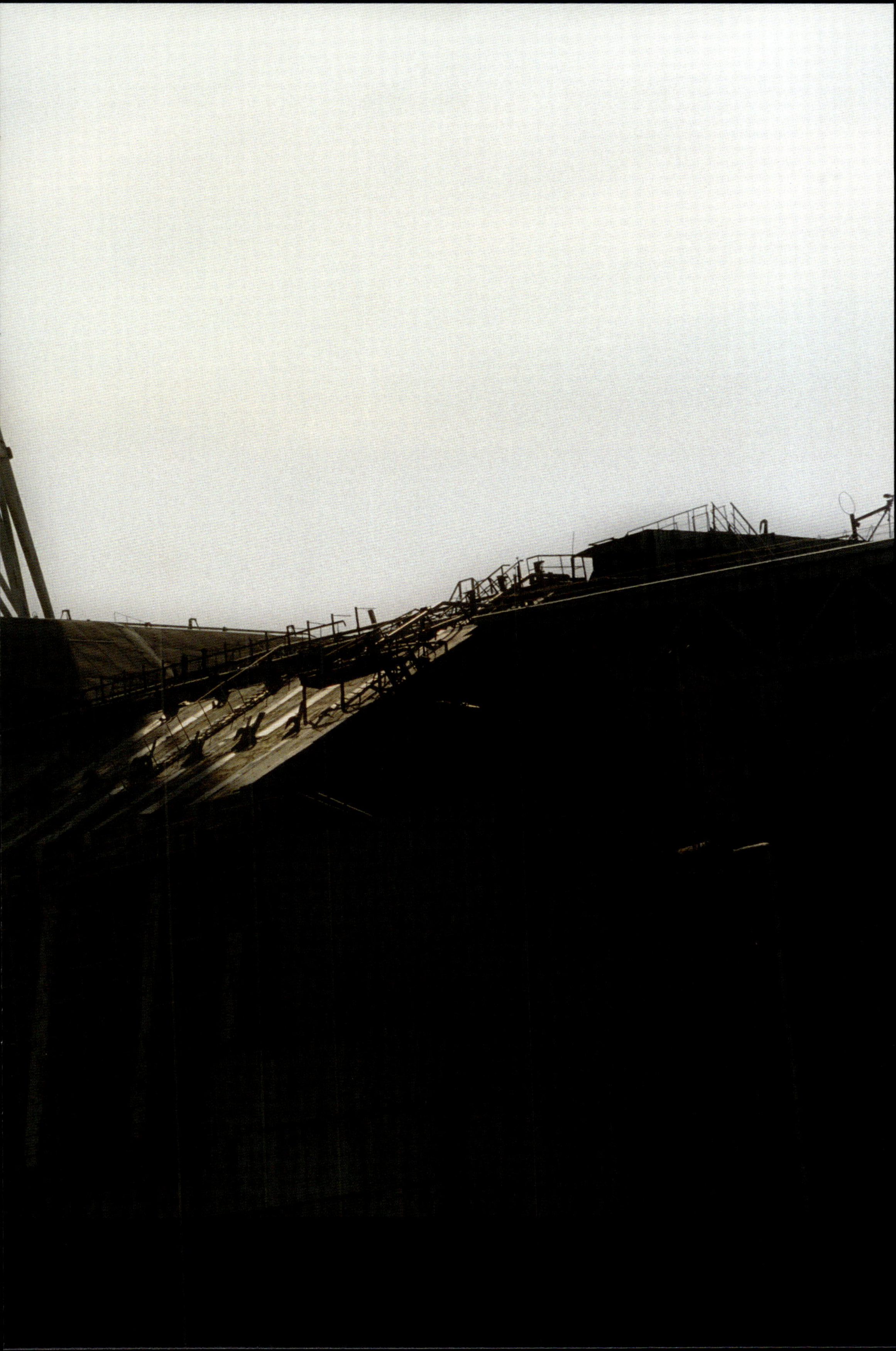

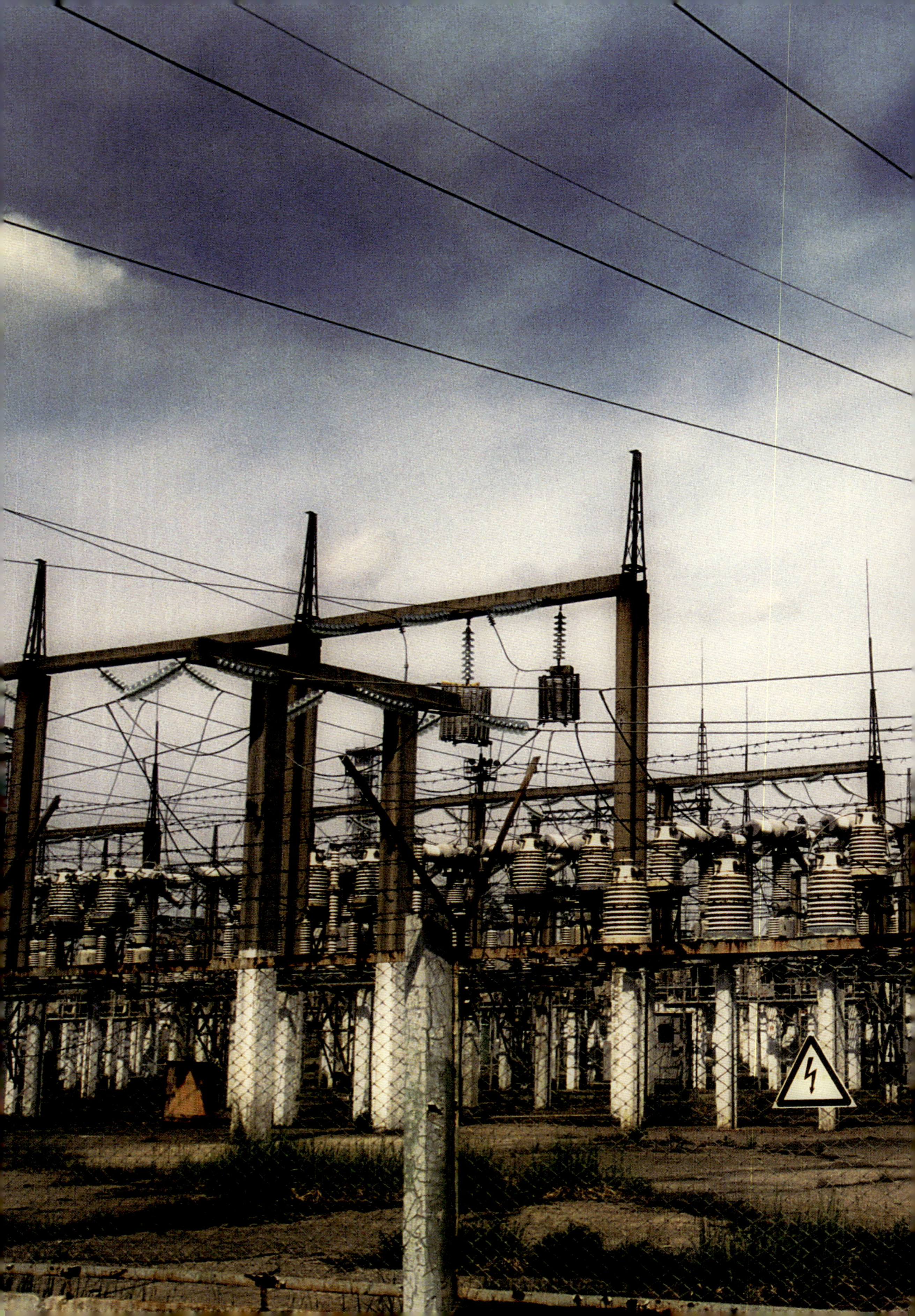

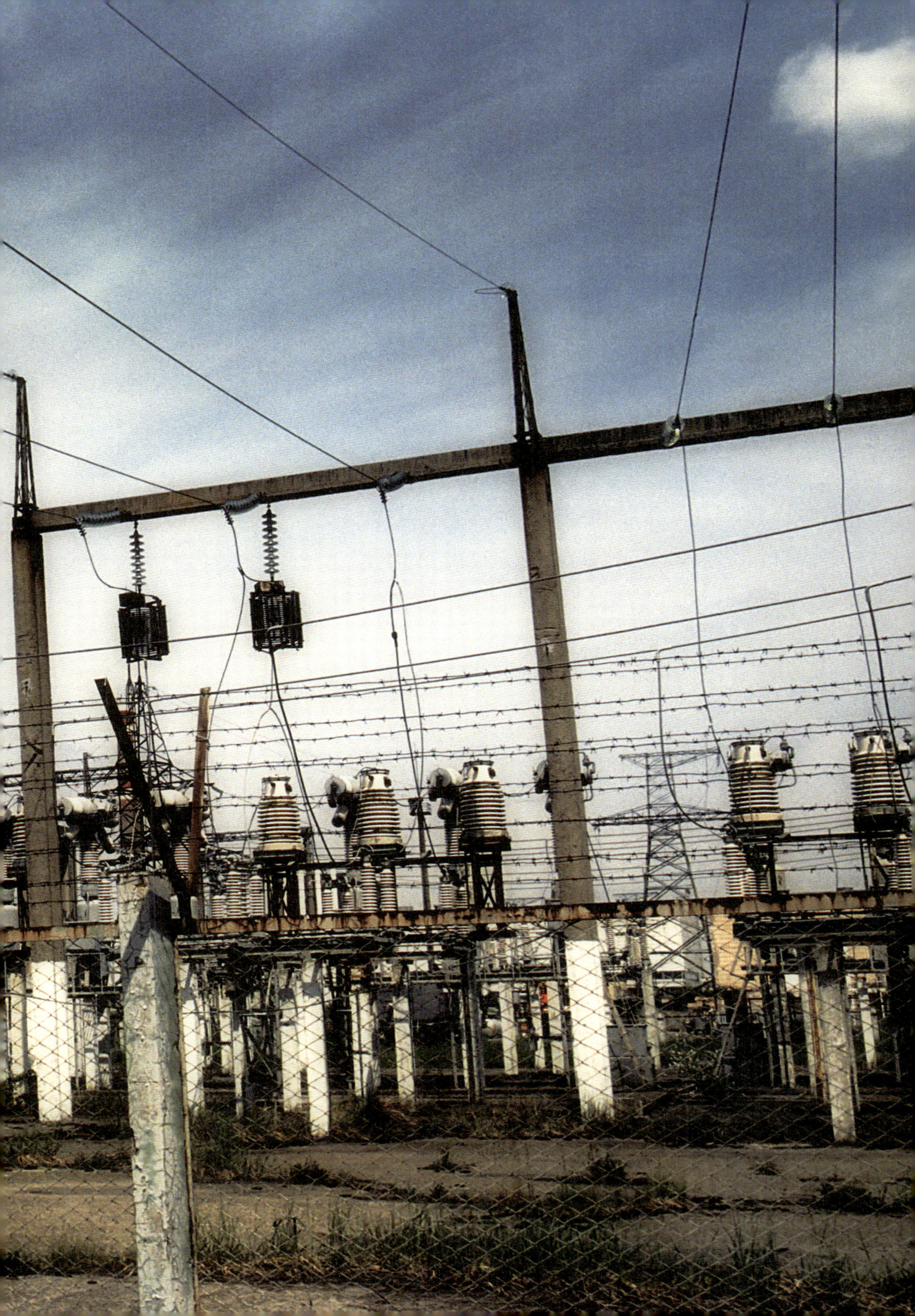

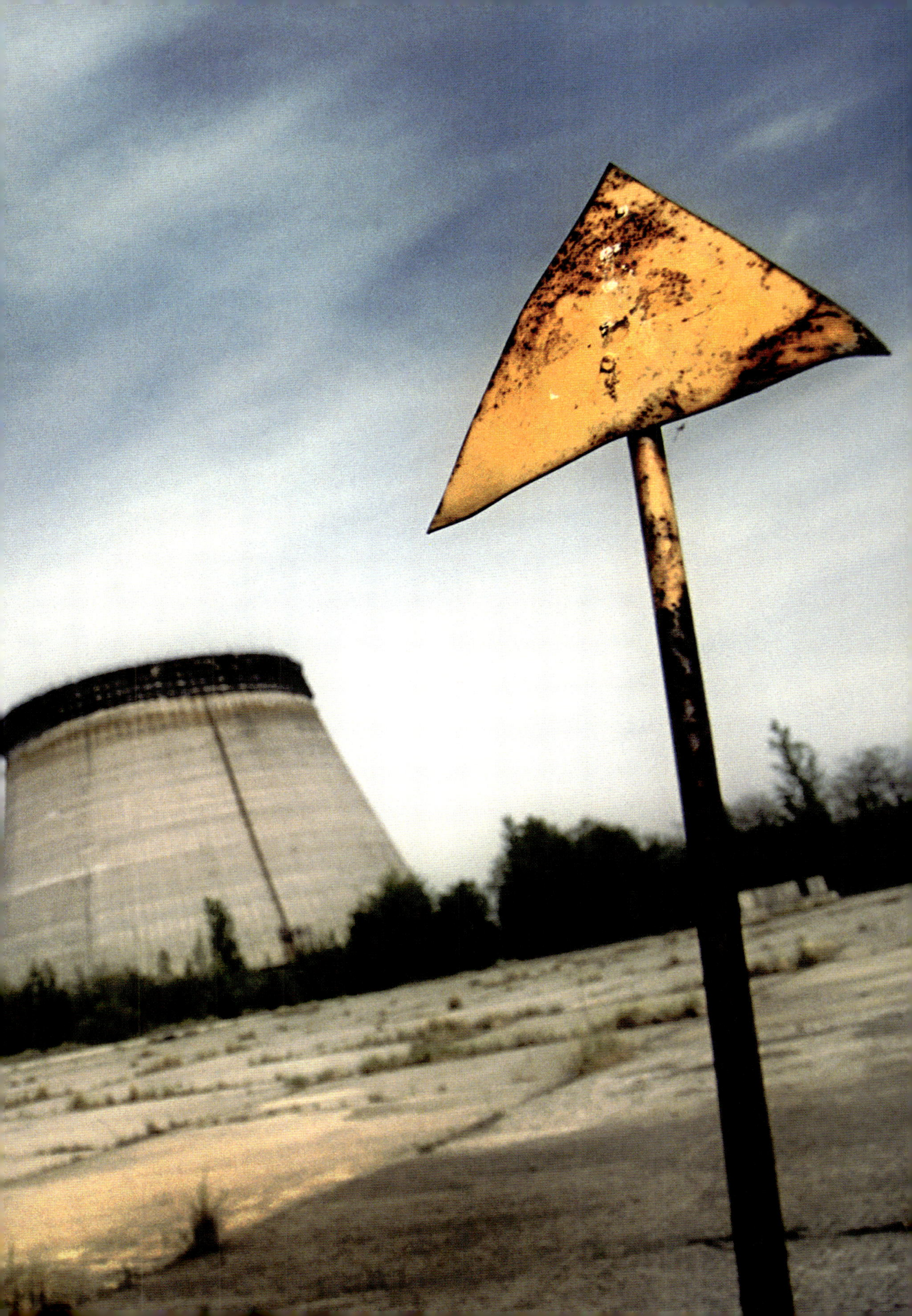

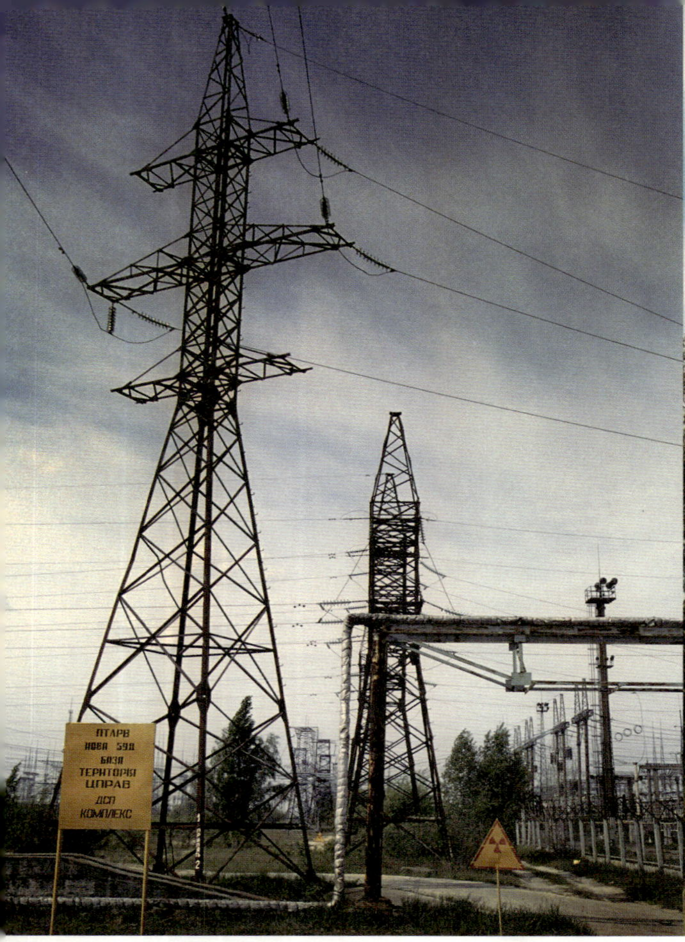
ПТАРВ
БАЗА
ТЕРИТОРІЯ
ЦПРАВ
ДСП
КОМПЛЕКС

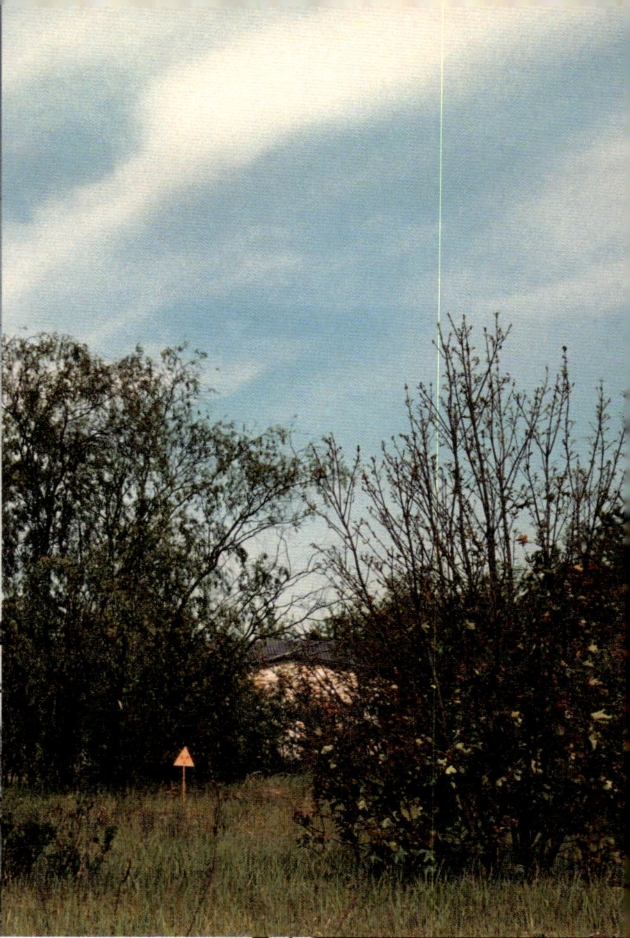

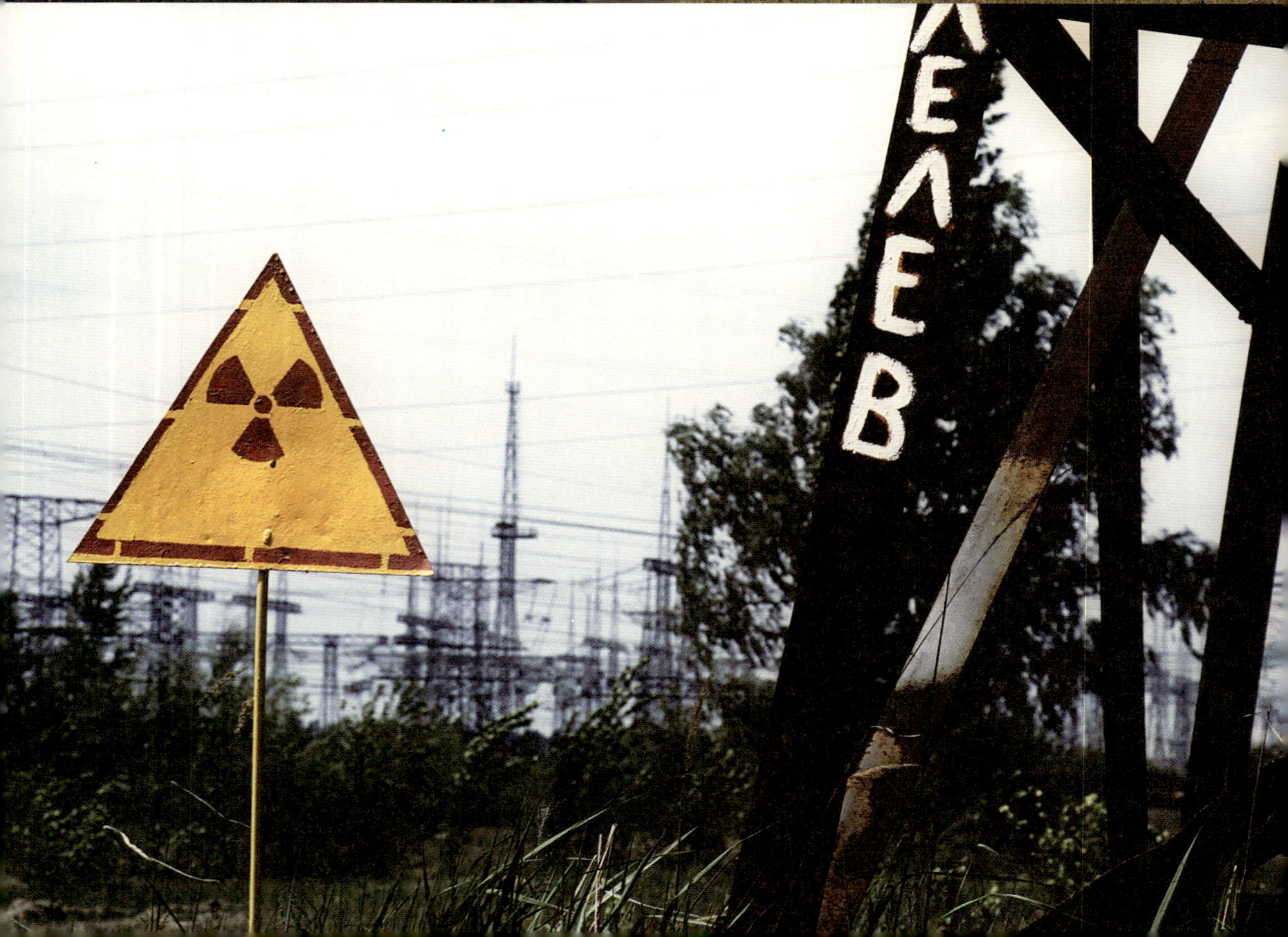
ЕЛЕВ

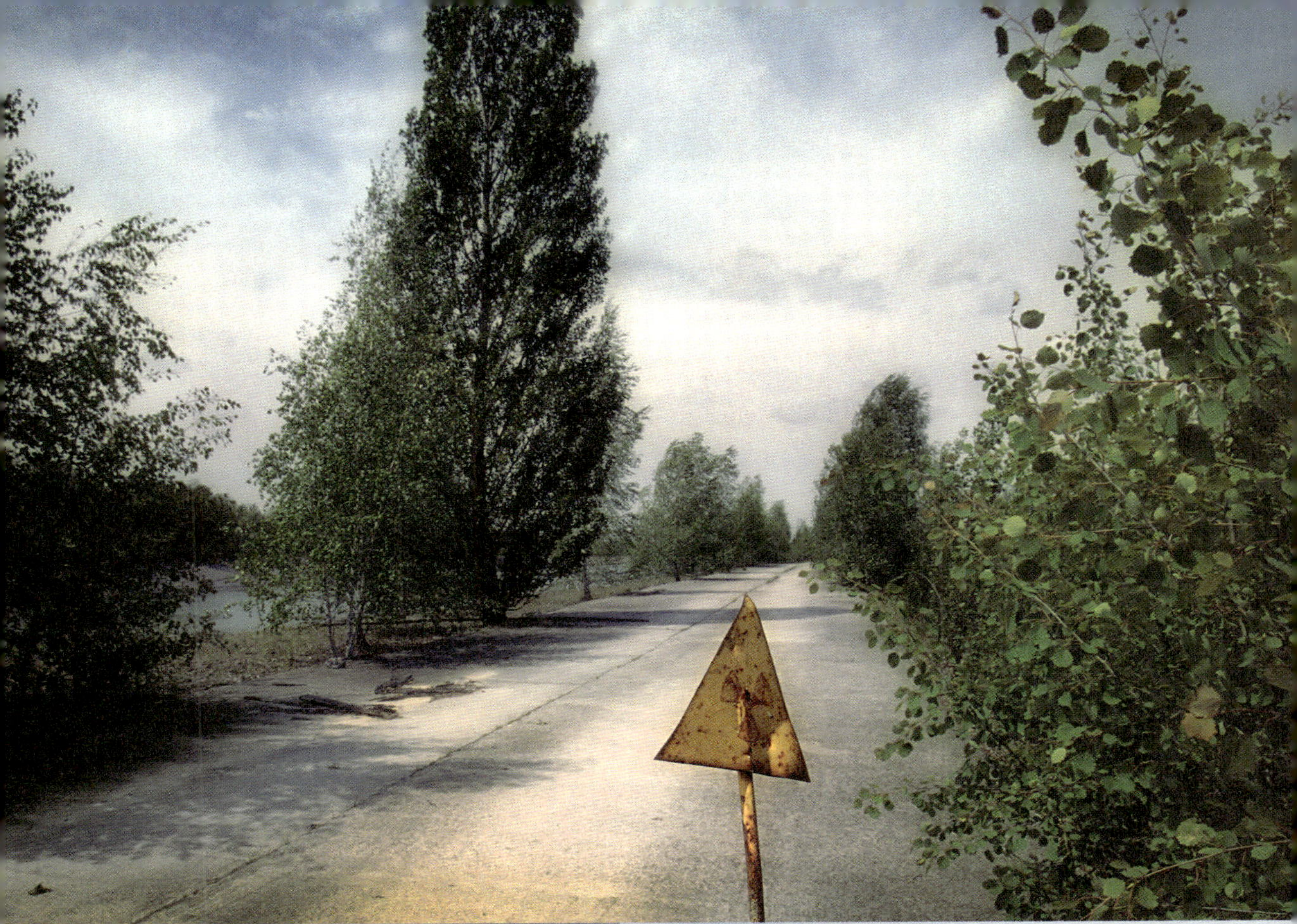

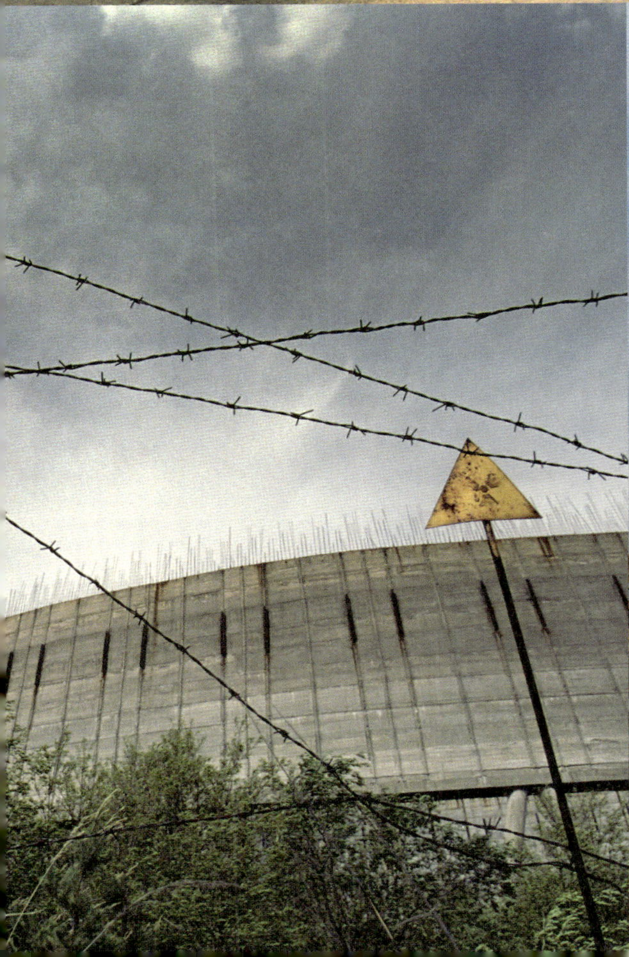

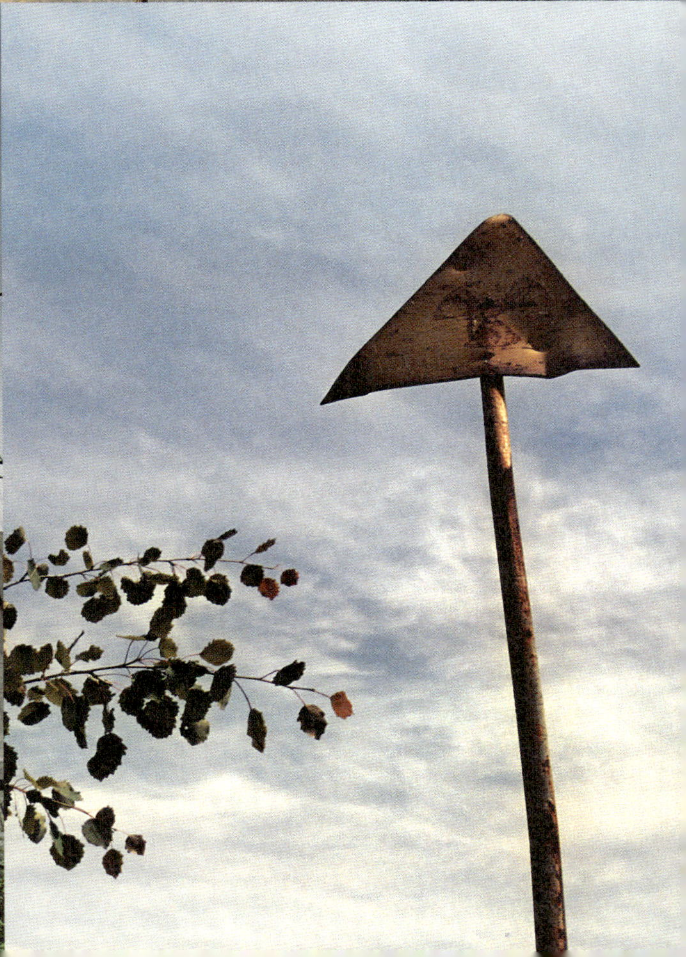

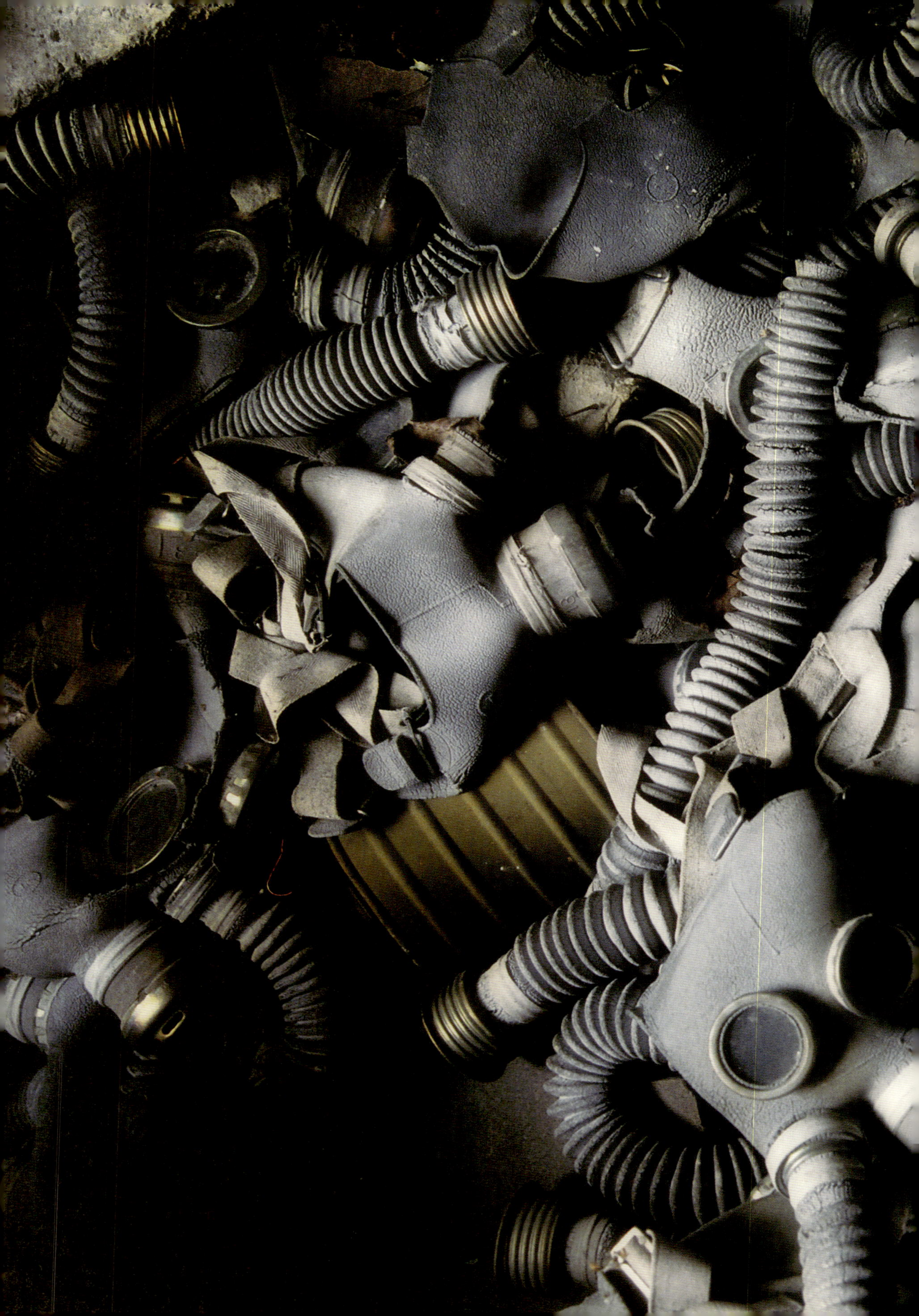

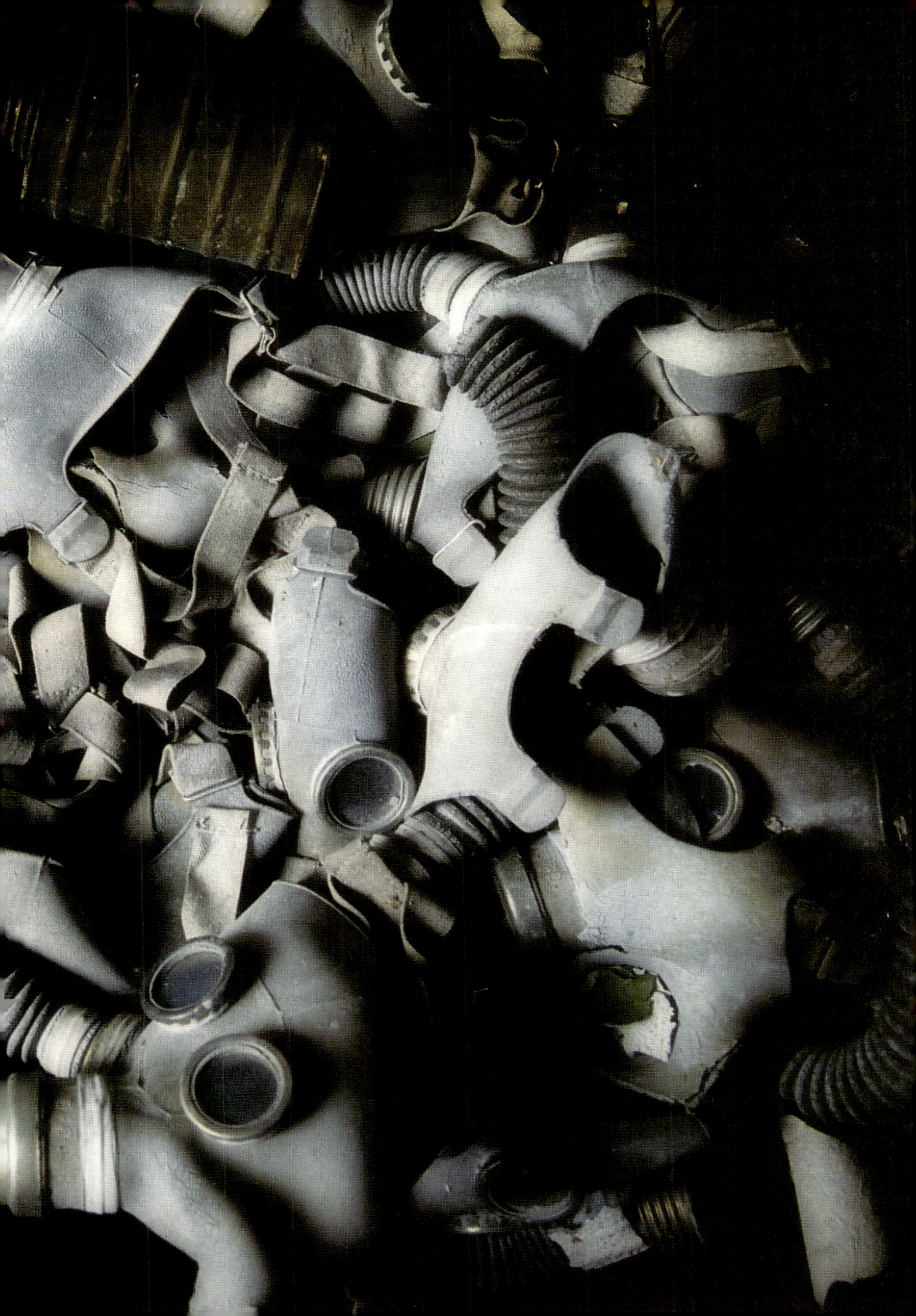

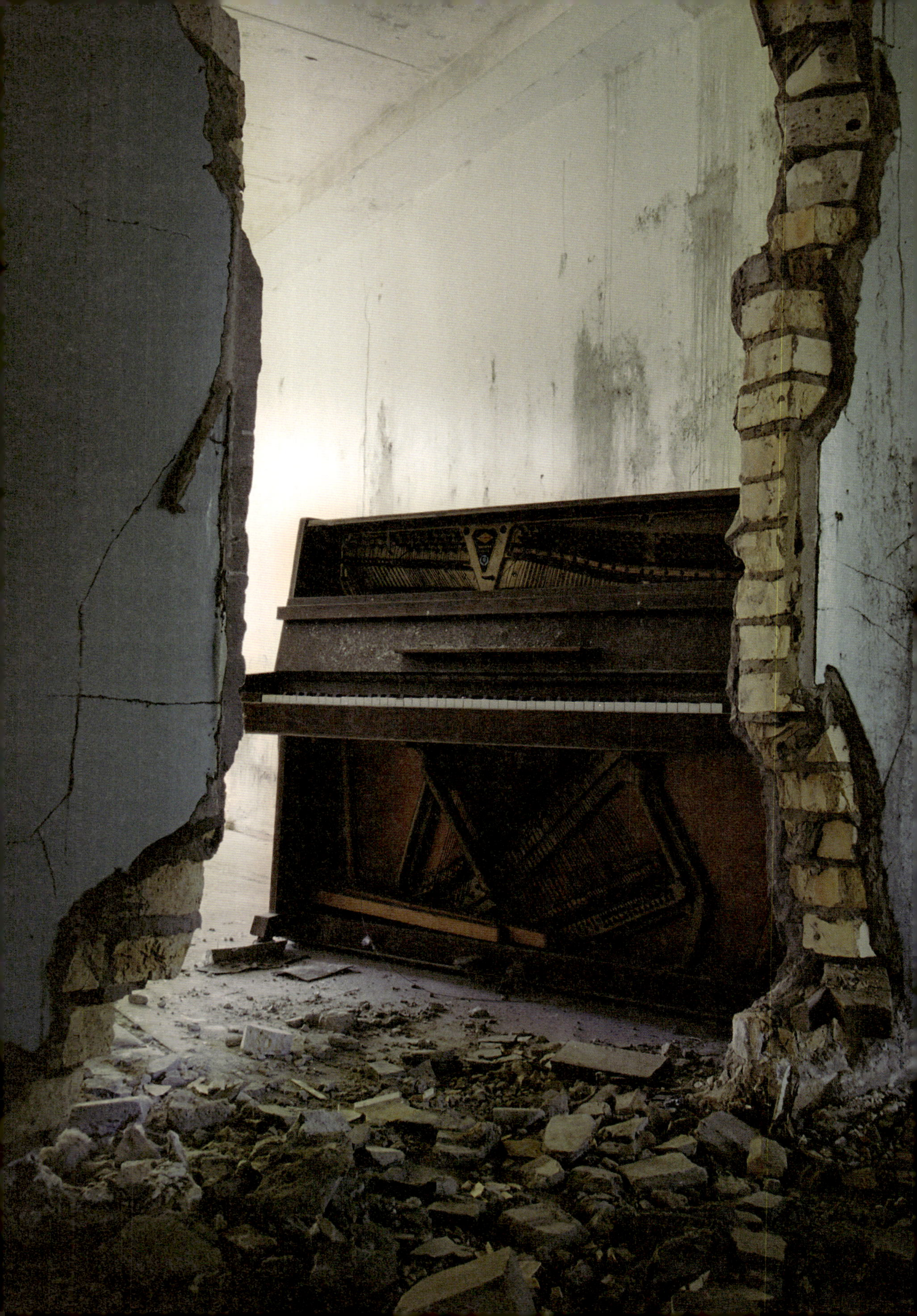

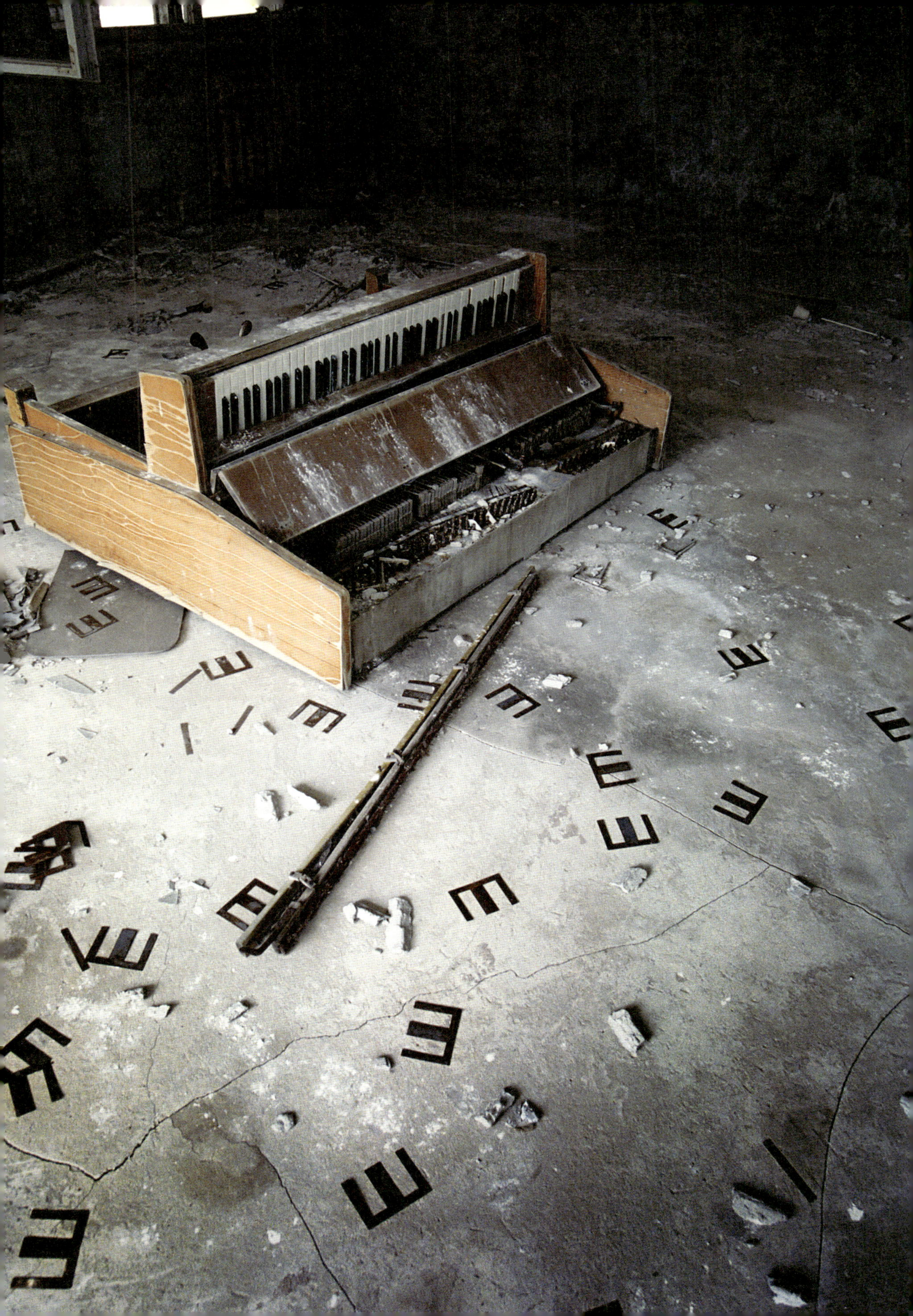

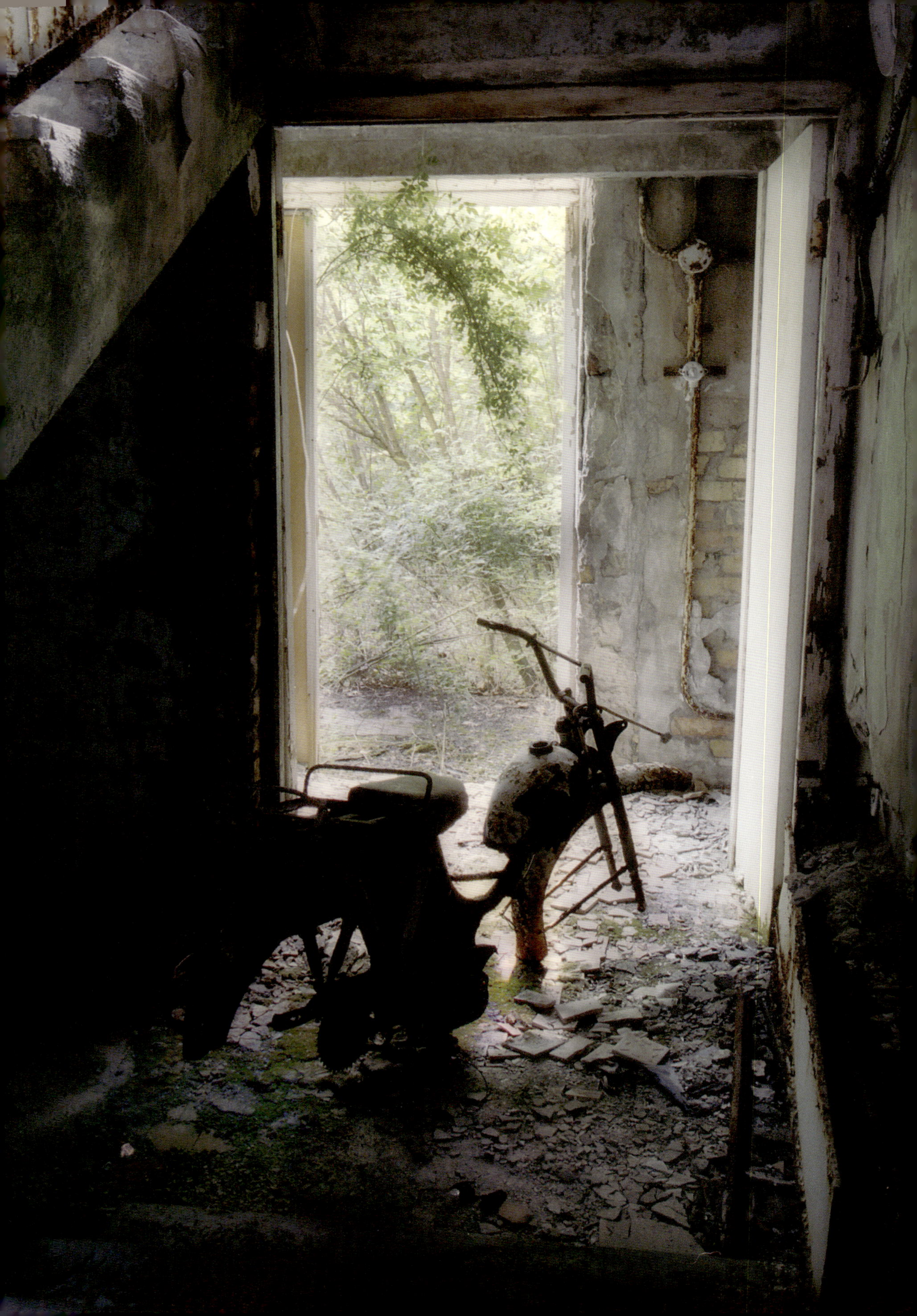

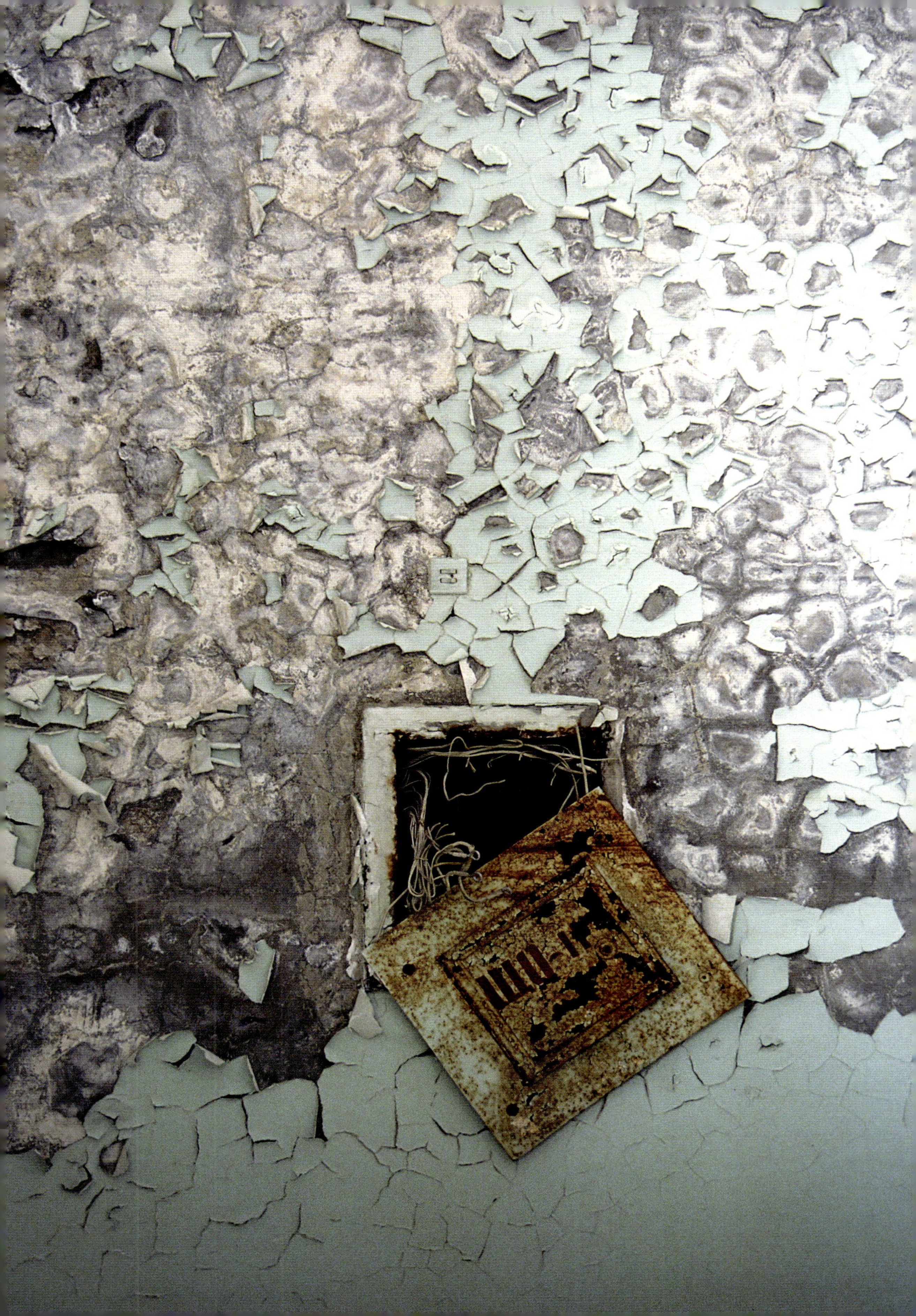

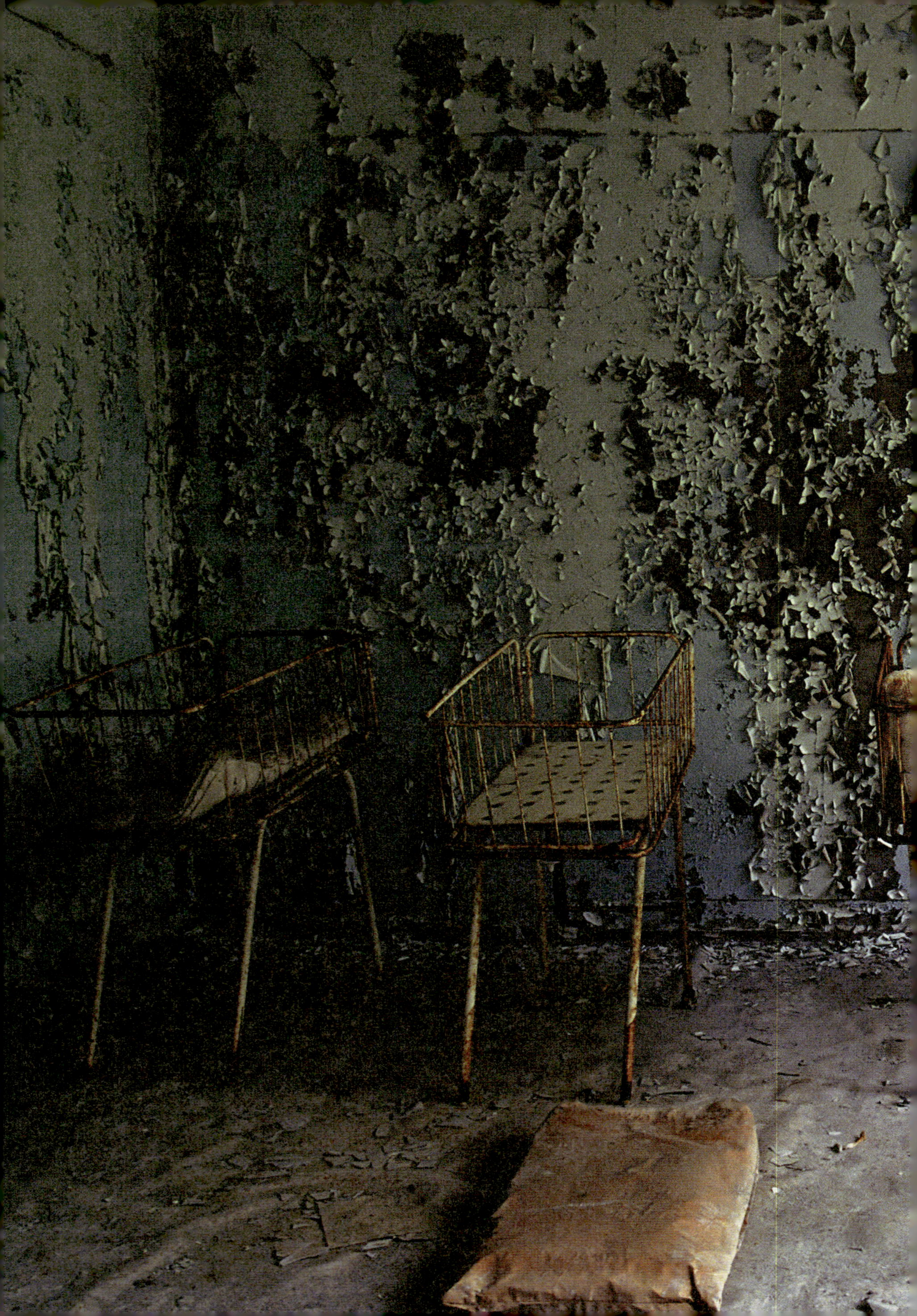

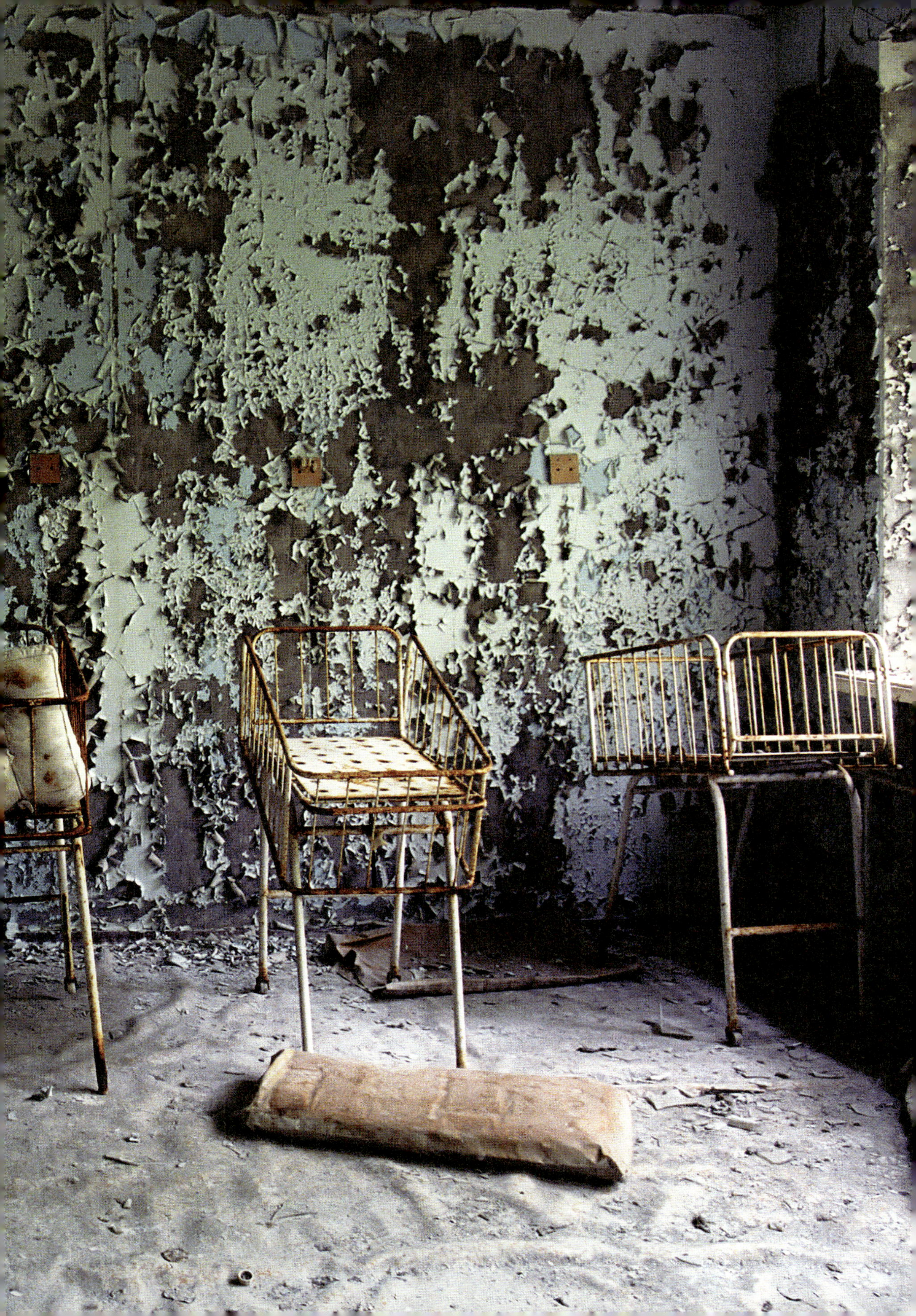

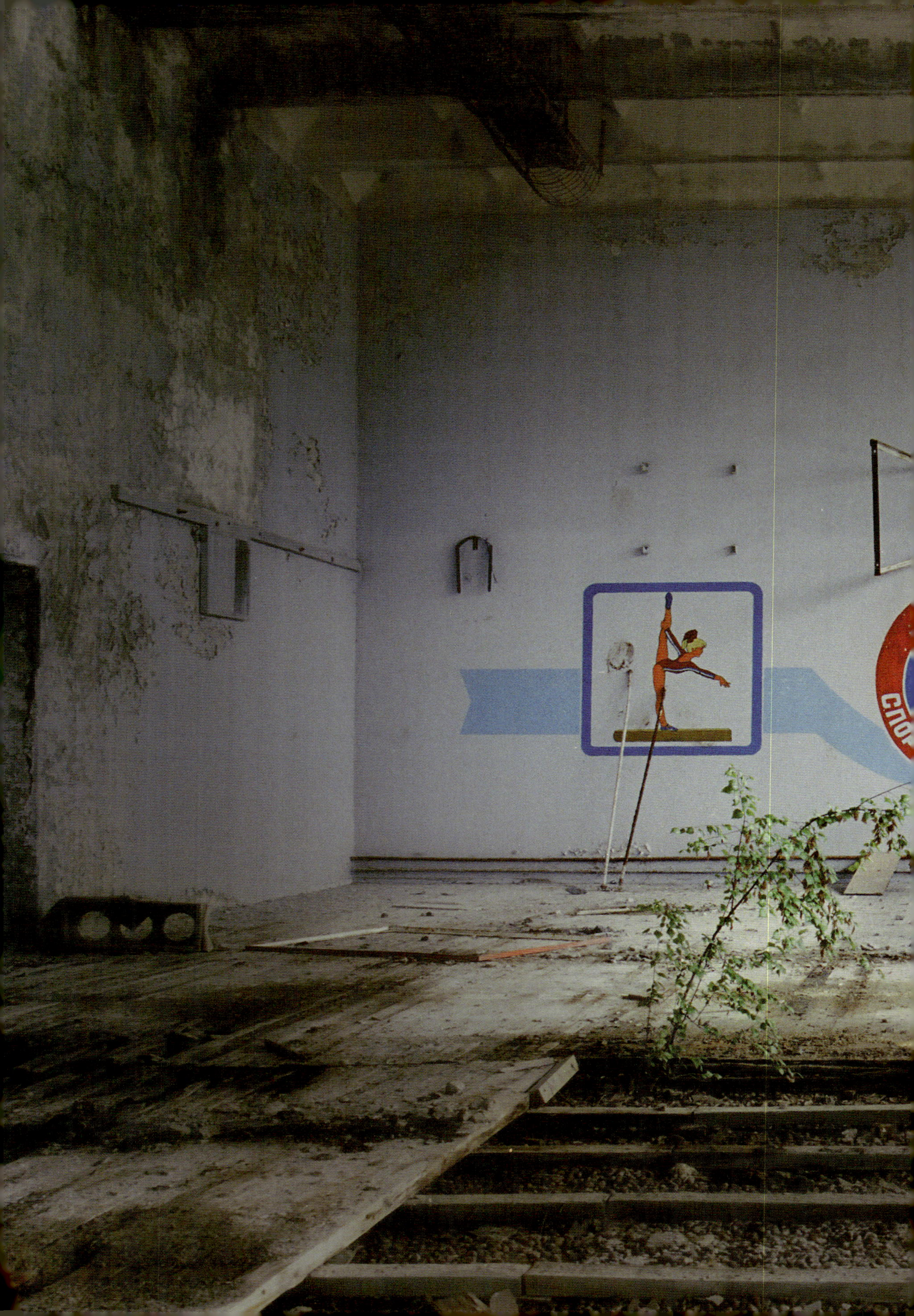

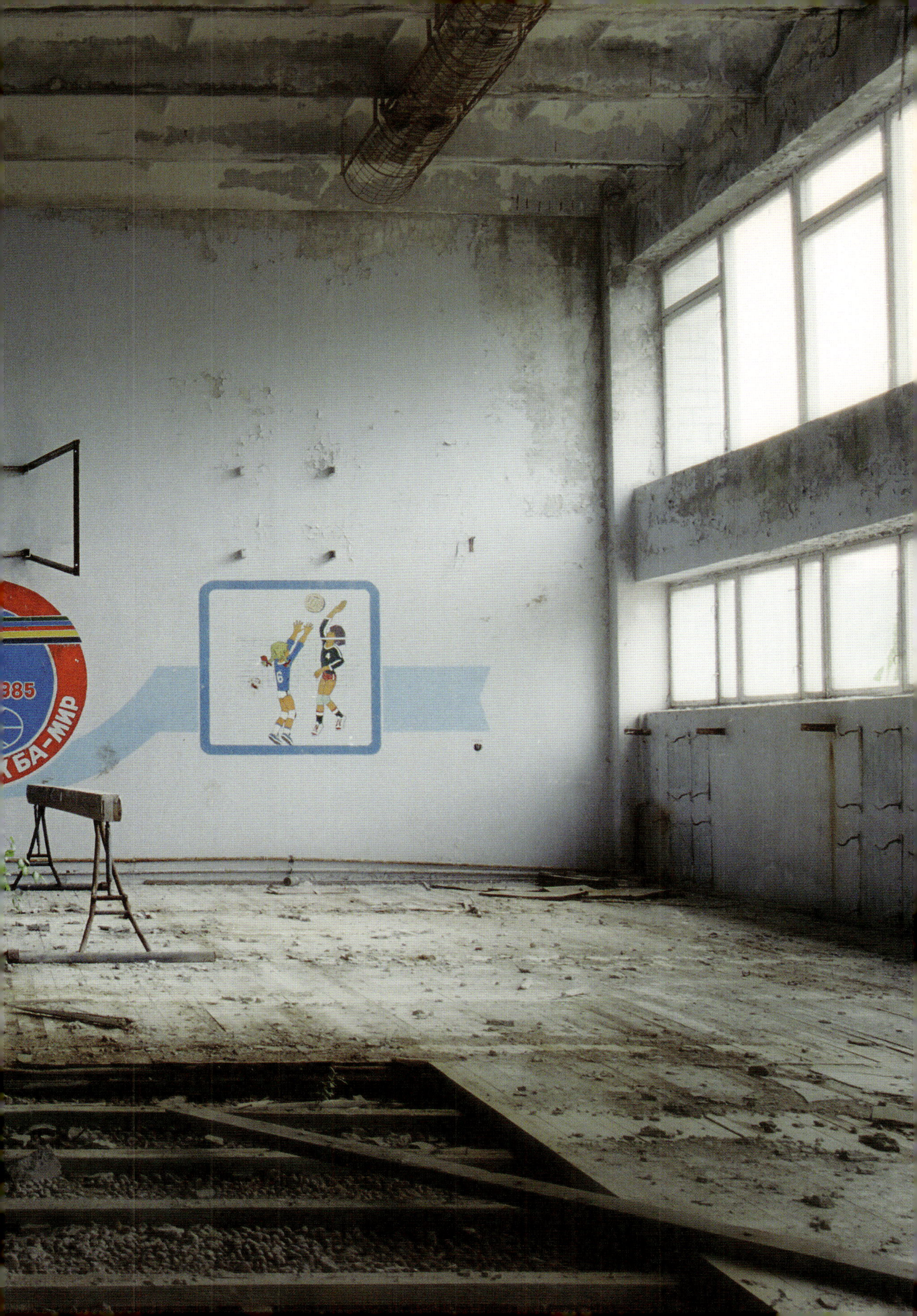

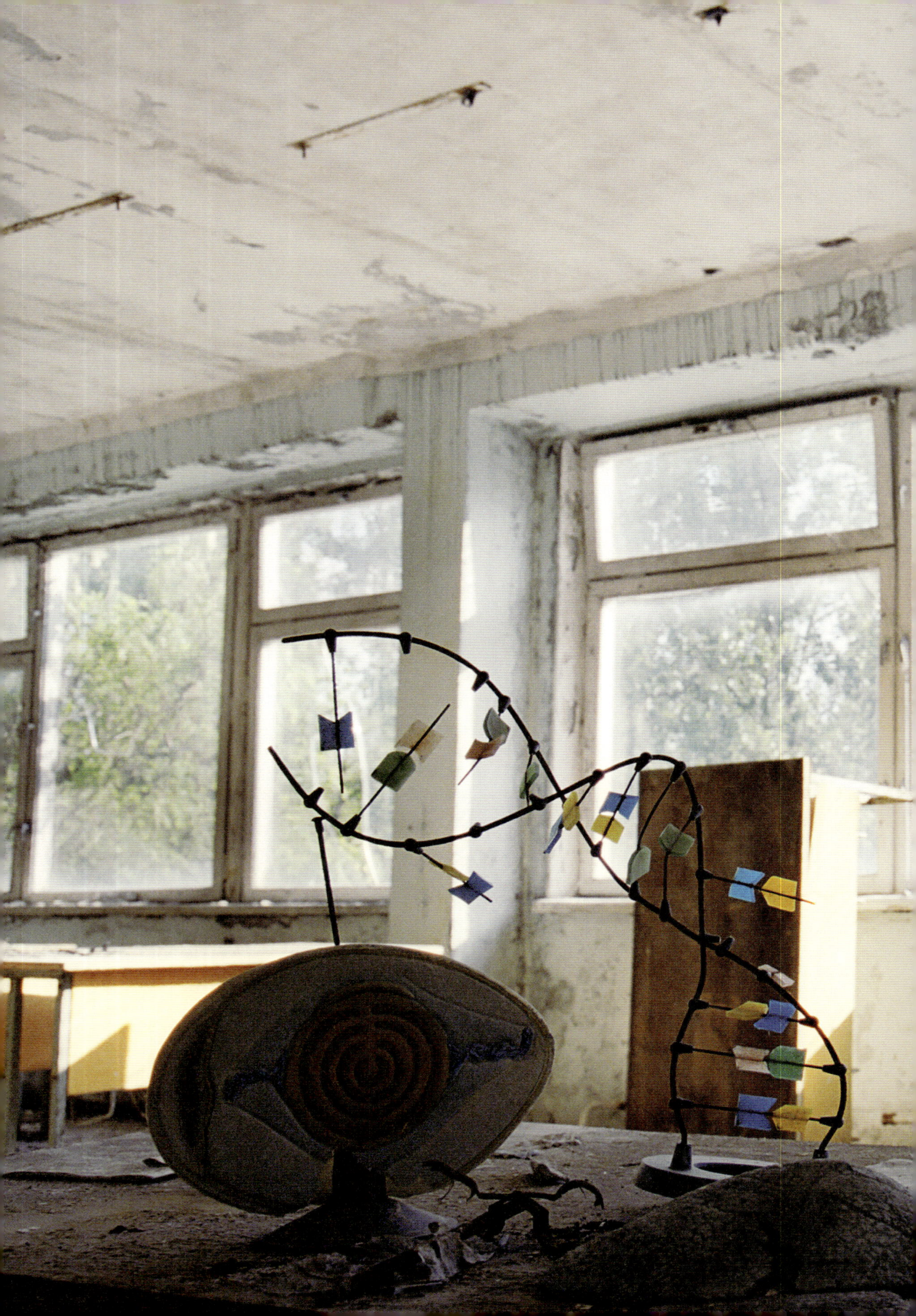

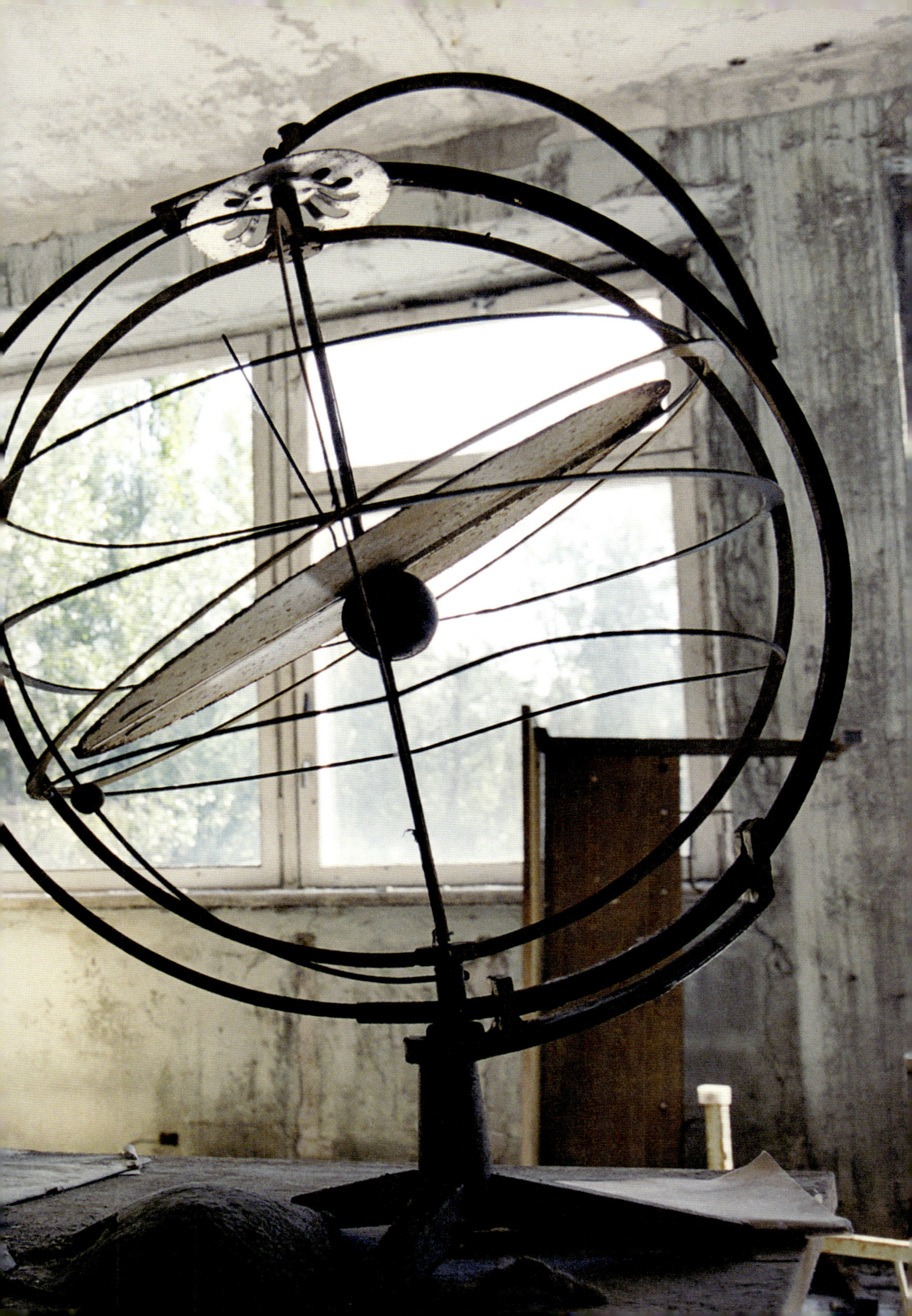

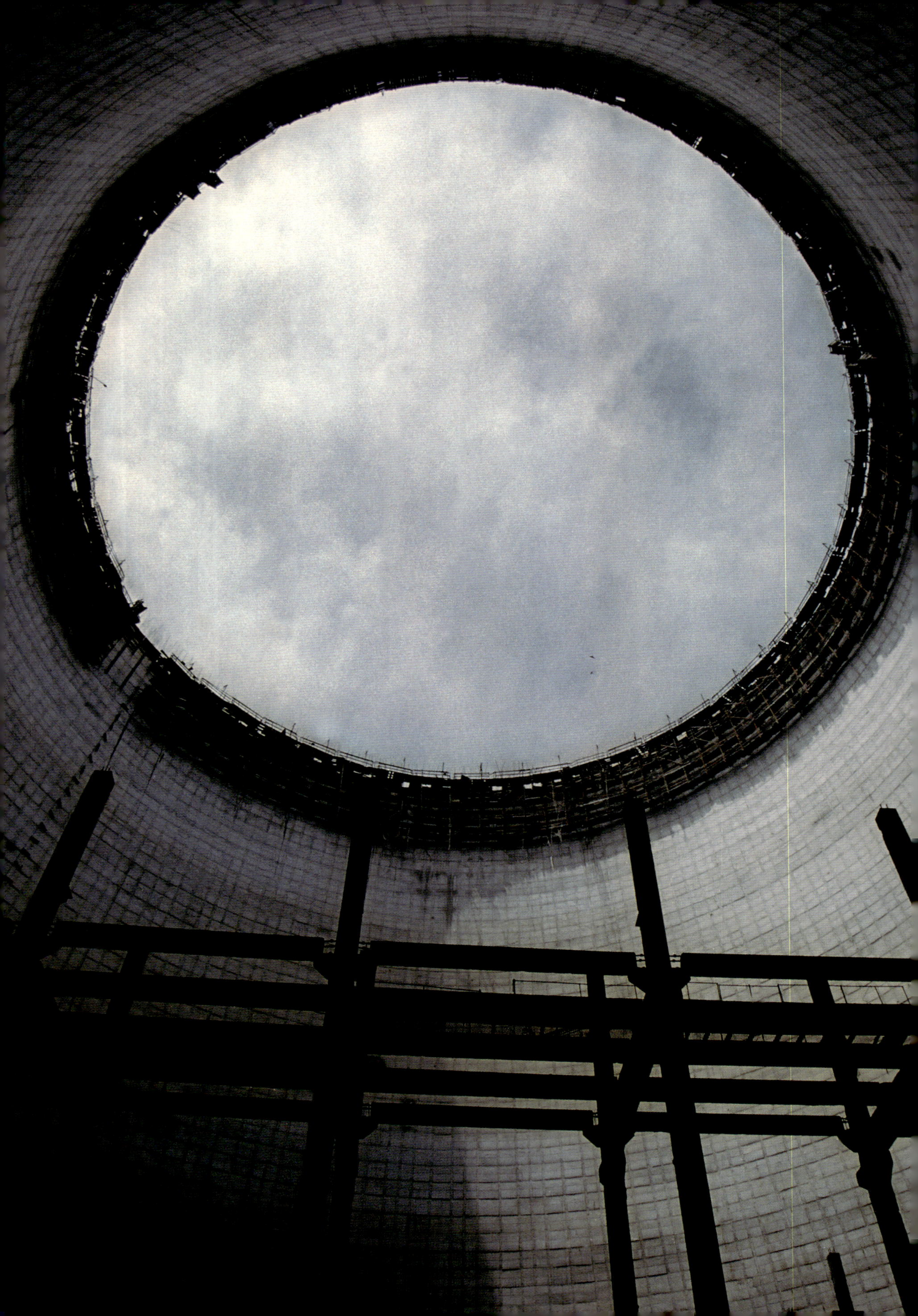

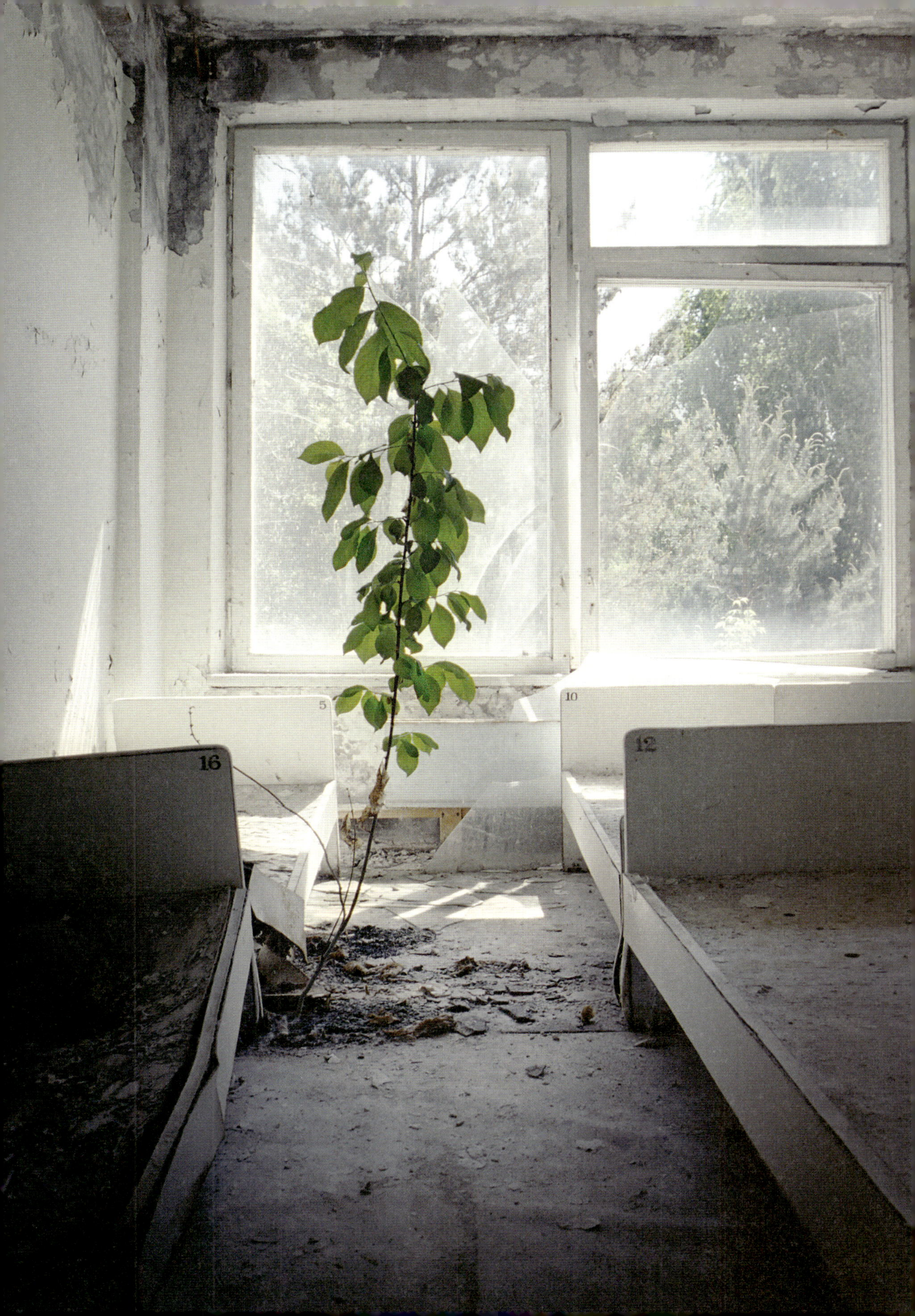
16
5
10
12

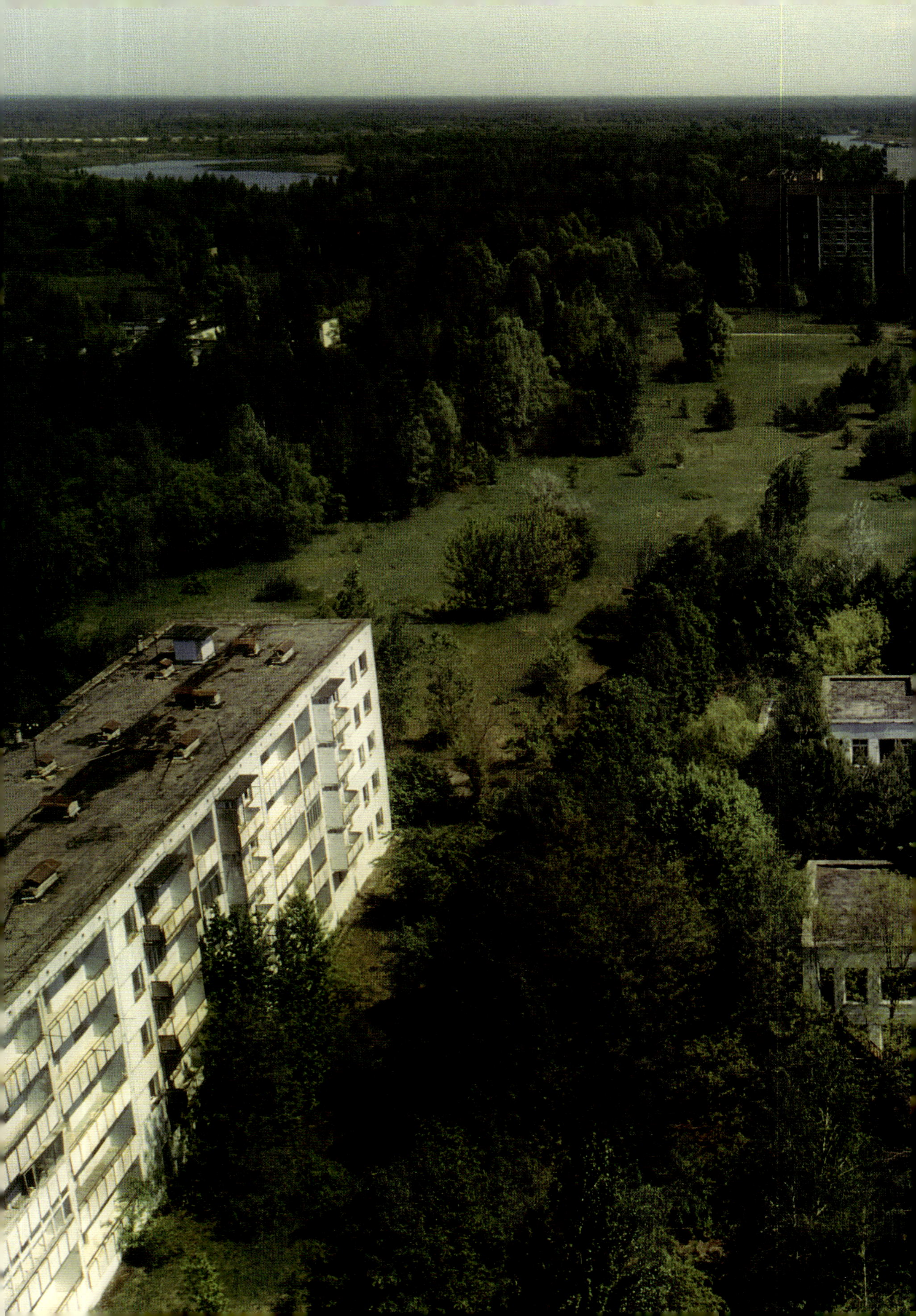

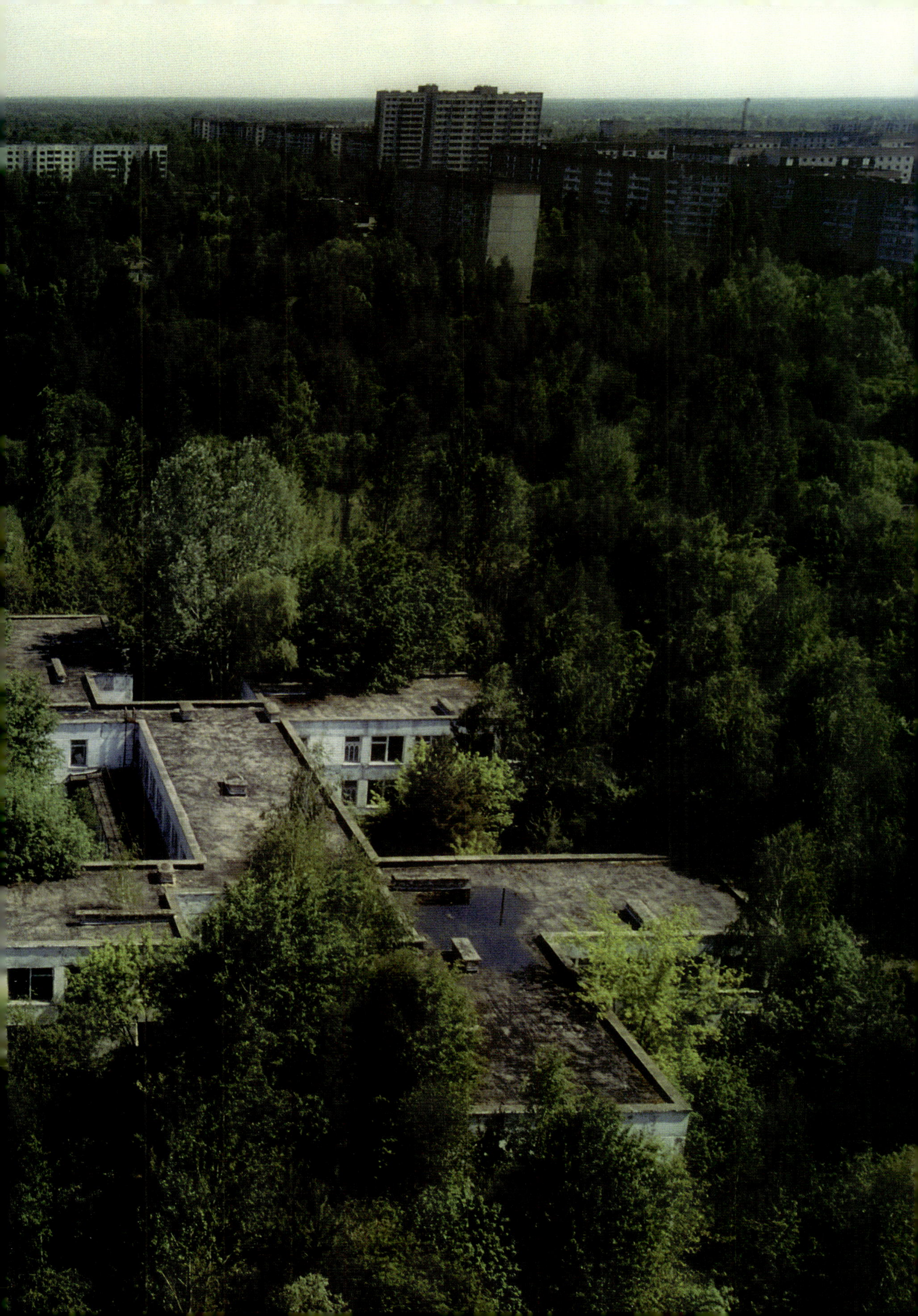

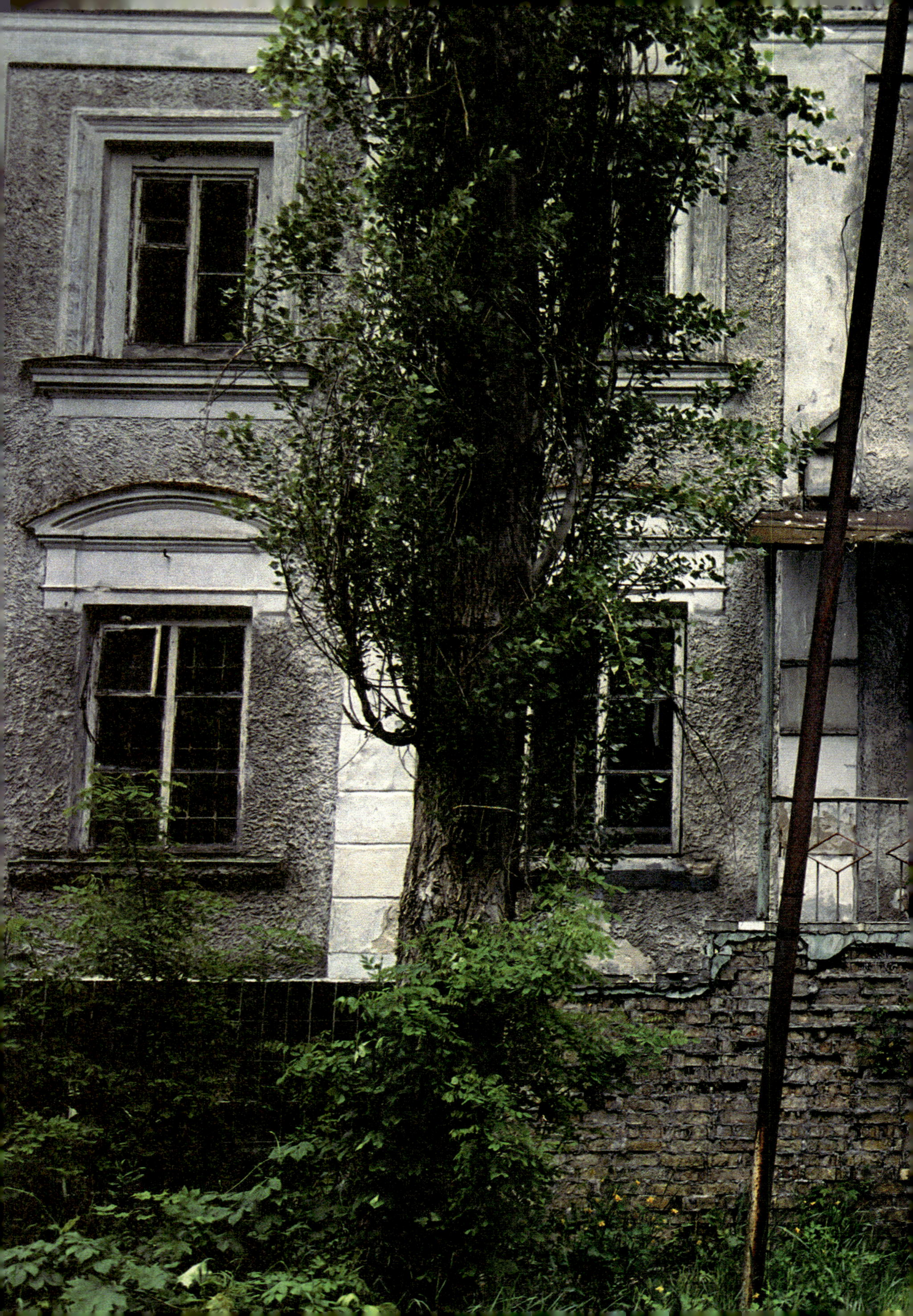

26

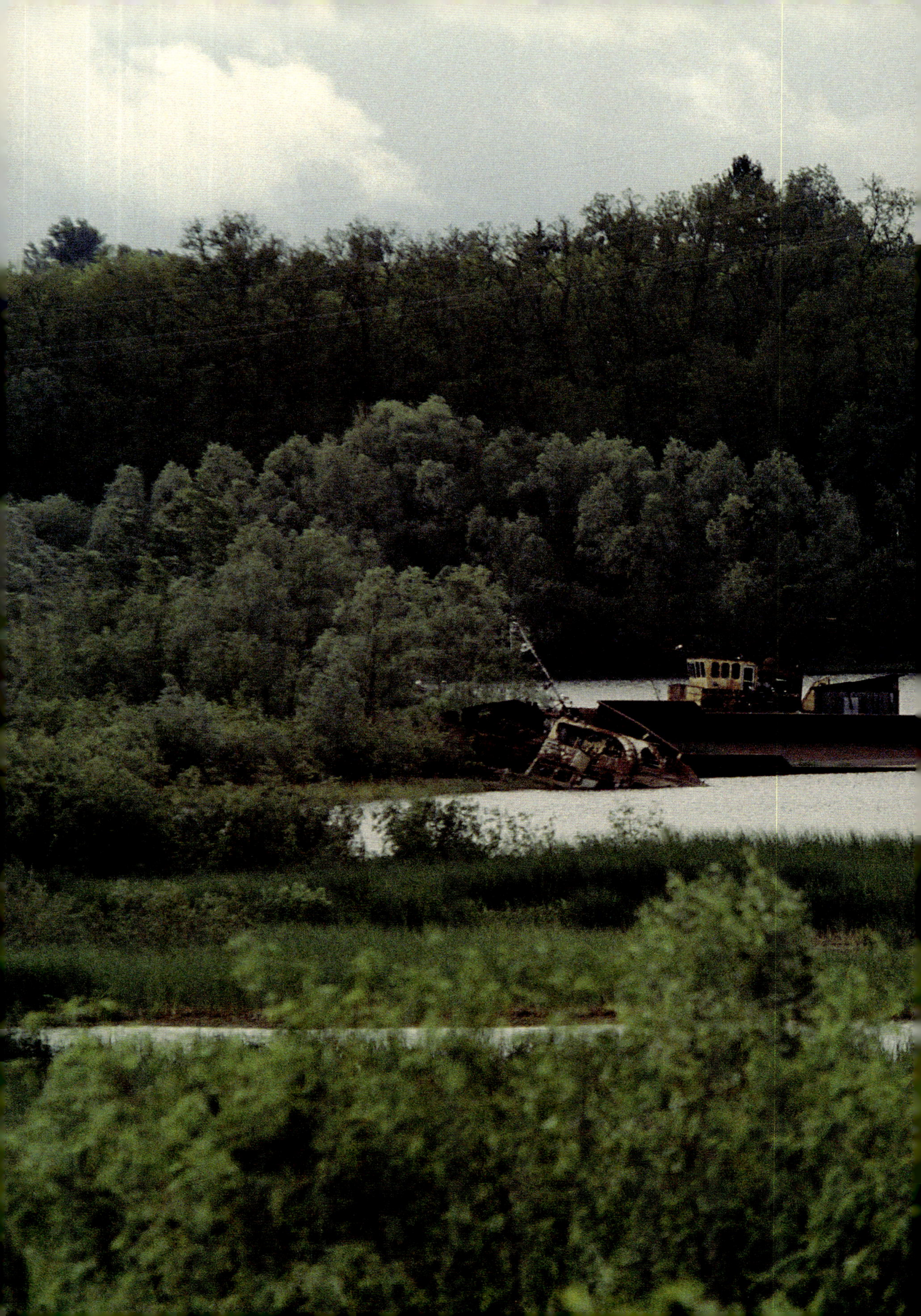

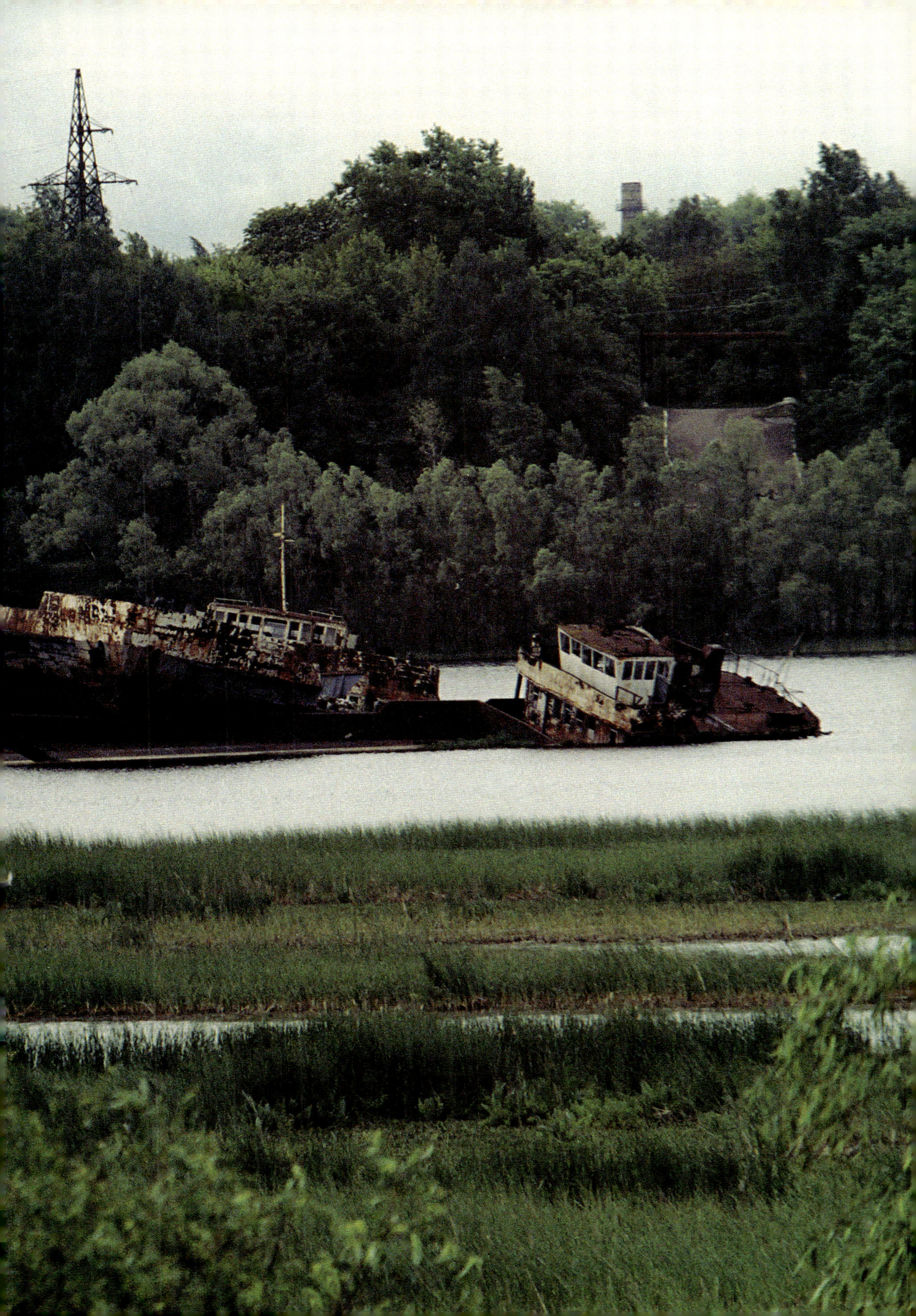

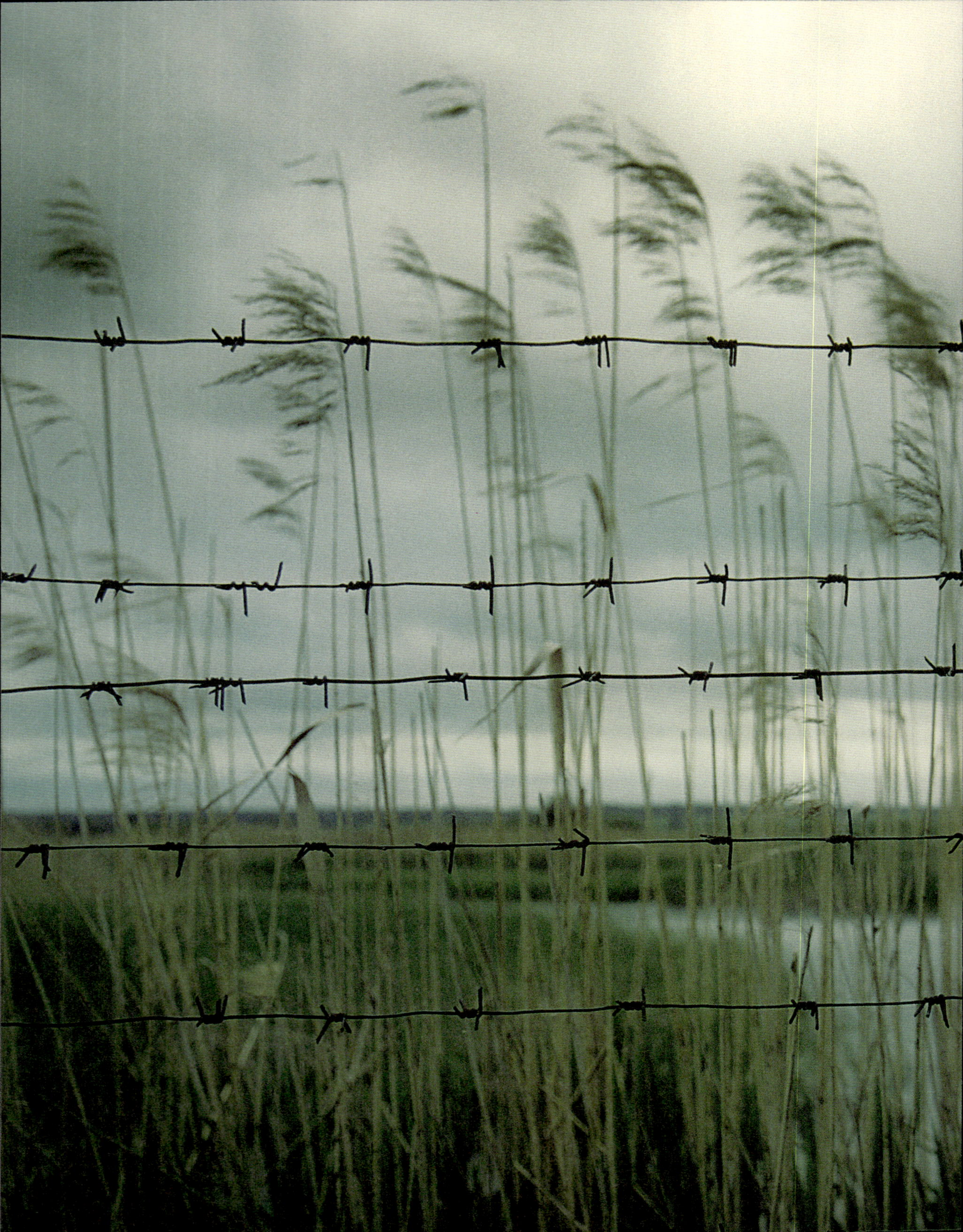

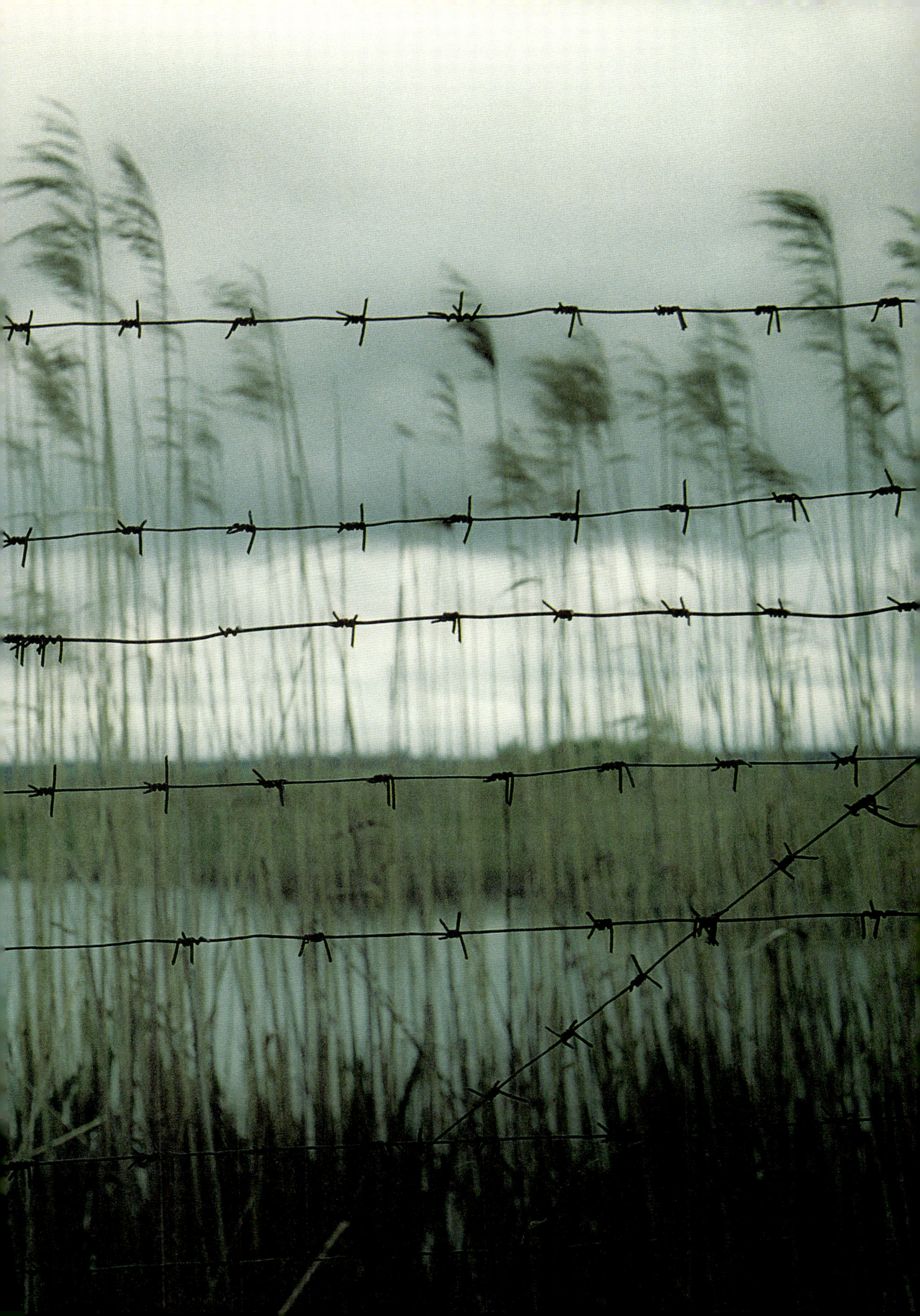

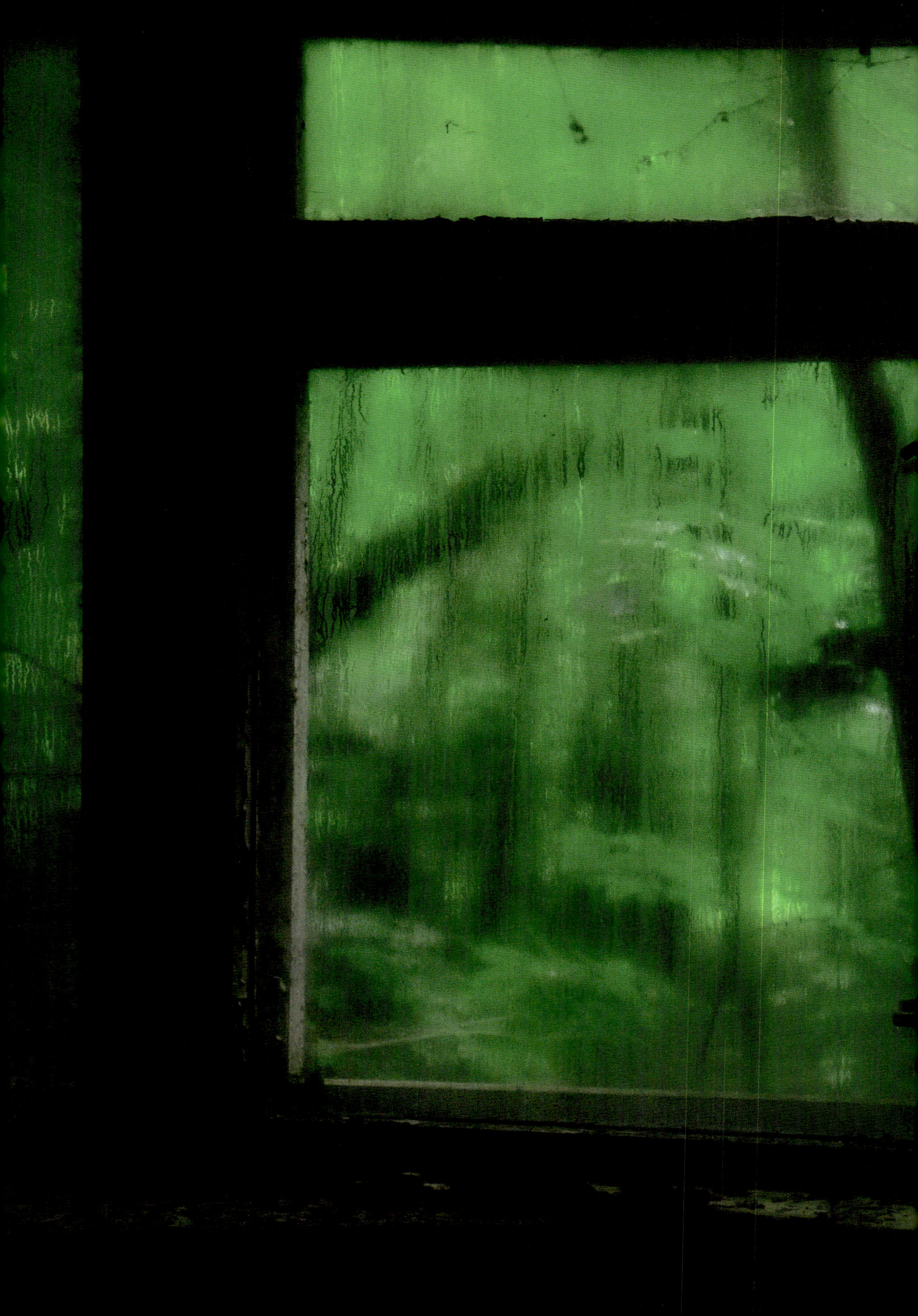

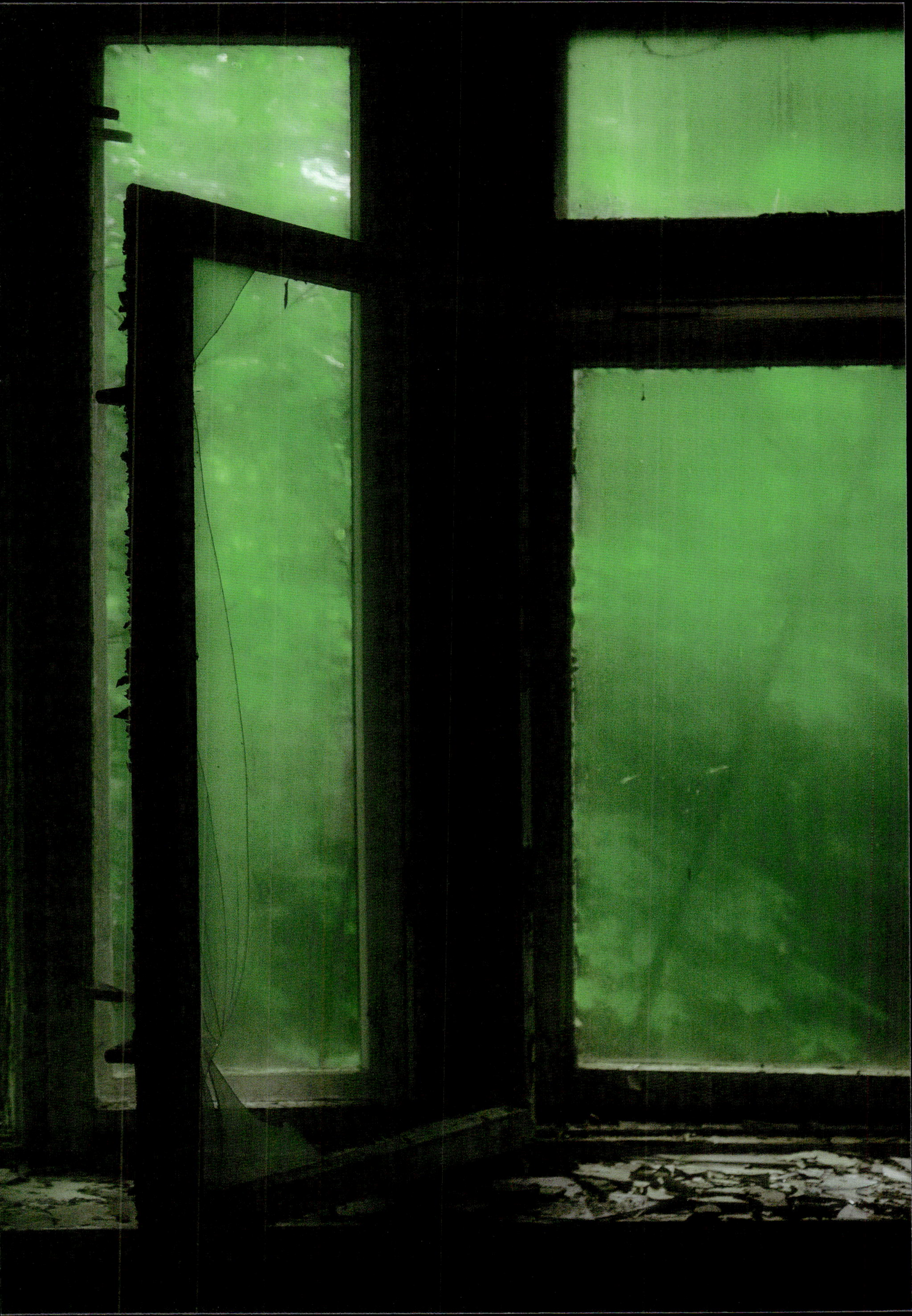

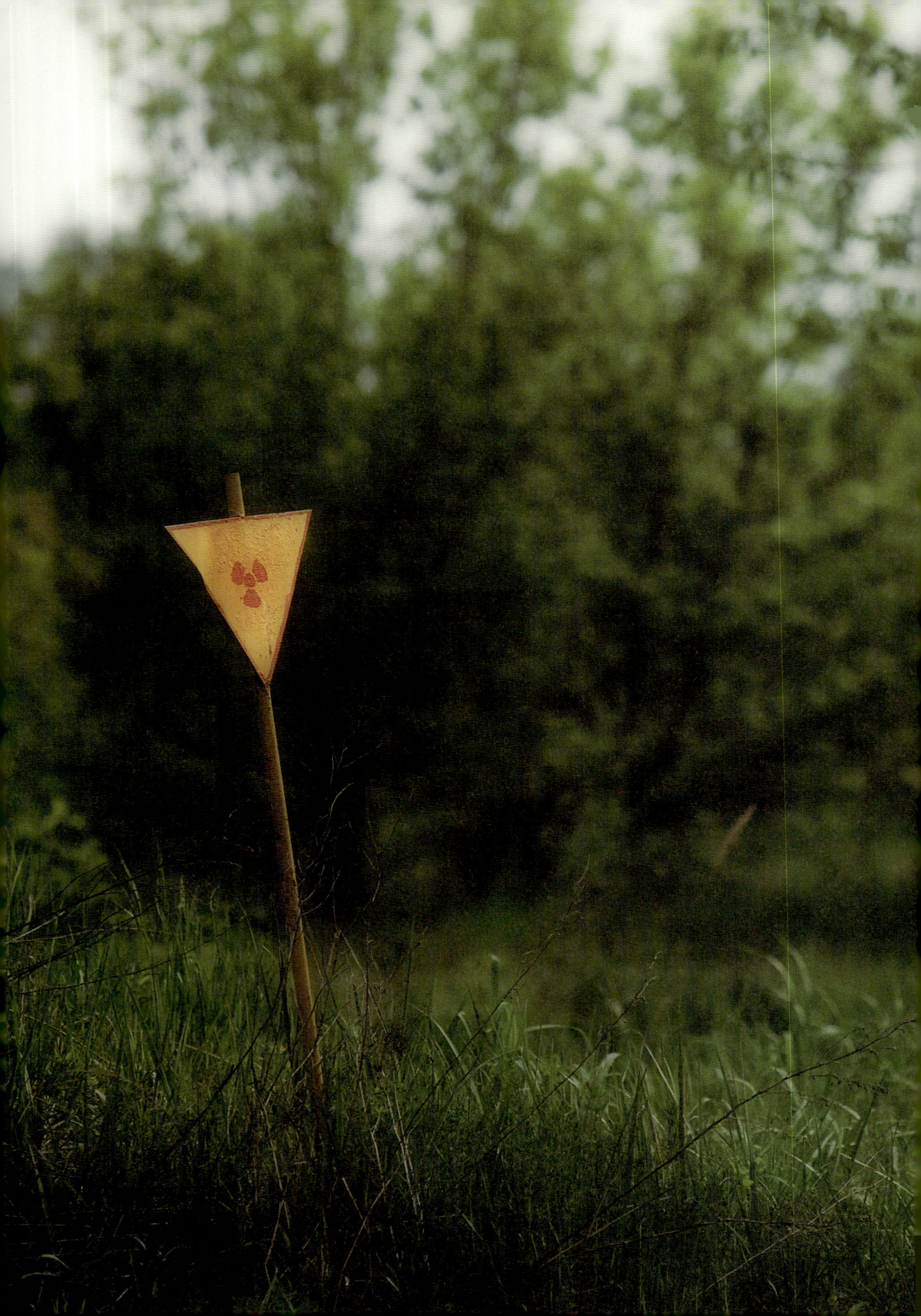

探訪チェルノブイリ

Rambling around
the abandoned land
of Chernobyl

ベラルーシ
ロシア
チェルノブイリ
ポーランド
キエフ
ウクライナ
ルーマニア
黒海

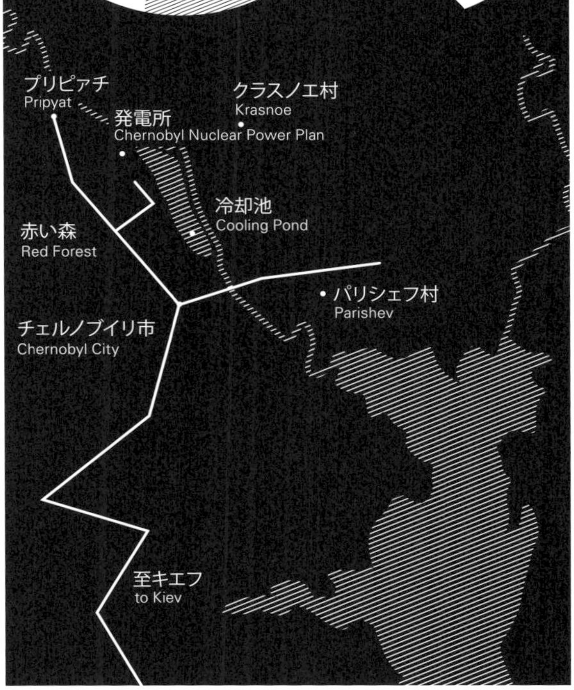

プリピアチ
Pripyat

クラスノエ村
Krasnoe

発電所
Chernobyl Nuclear Power Plan

冷却池
Cooling Pond

赤い森
Red Forest

パリシェフ村
Parishev

チェルノブイリ市
Chernobyl City

至キエフ
to Kiev

どこにいても春の空気は奥ゆかしい。可憐に咲く色とりどりの花の芳香も、ウグイス、ヒヨドリの合唱も柔らかい春風に吹かれ、廃墟になった16階建てのマンションの屋上にいる僕らのもとまで届いて来た。朽ちた塵埃だらけの階段を息急きながら上って来た僕らへのねぎらいの拍手のようでもある。
　チェルノブイリ原子力発電所の城下町プリピアチ。1986年の発電所事故の後5万人いた住人は即座に強制退去させられ、それ以来ポレーシェと呼ばれるドニエプル川が作った大低地帯にその亡骸をさらし続けている。地面をさまよっているとなかなか気付かないが、こうしてアイポイントを上げると発電所という城を中心にして北側へ放射状に奥深く広がっているのが手に取るように分かる。果てしなく続く広葉樹の森に窓ガラスが割れ壁のはがれかけたマンションがポツリポツリと建っている様はまるで荒れ果てた墓場のようだ。高度な科学技術がもたらした負の側面を垣間見る文明の抜け殻とでもいおうか。そんな街の俯瞰図に一瞬心が痛むが、目に飛び込む木々の緑は若々しく瑞々しい。
「ここの景色はすばらしいですね」
　通訳のアレクセイは、長年プリピアチを訪問しているがこの場所に来たのは初めてのようだ。
　チェルノブイリ原子力発電所を中心に半径30キロの区域は通称「ゾーン」と呼ばれている。発電所事故の後にばらまかれた放射能の影響で今も赤錆びた鉄条網で囲まれた禁断の地だ。ゴーストタウンと呼ばれる場所は世界にも数々存在するが、ここは核汚染されているという点で特別な存在だ。
「純さん何度も言っていますが膝をつかないでください」
　ばらまかれた放射能は忍者のように鳴りを潜めていて、計測器がないとその存在は認識することができない。ちょっとした狭いスペースがホットスポット（高濃度汚染地）になっていて、カメラマンの癖でもある膝付きでの構えは、被曝の可能性があるそうなのだ。
「じゃ、三脚はどうなんだ？　コイツは既に土にまみれているが……？」
「捨ててくださいね。チェルノブイリの悪いお土産です」

　しかし度々の忠告も新緑の緑の森の美しさとそこに佇む廃墟の威厳に感嘆していると頭のどこかに行ってしまうものだ。そして放射能は自分の存在を僕らにアピールすることもない。
「今日は切り上げて街に戻りましょう」
0.5ミリレントゲンと表示されたガイガーカウンターを指差してアレクセイは夢中になってシャッターを切っている僕の注意を喚起する。僕らは春のチェルノブイリの果てしない緑の海をもう一度目に焼き付けてから朽ちた階段を小走りに下った。

小さな街チェルノブイリ

　チェルノブイリ市は発電所から南に10数キロ下ったところにある小さな街だ。春の芽吹きが一段落ついた頃、街にはセイヨウトチノキの白い花が咲き乱れていた。プリピアチと比べ小さな街だが、事故当時には教会があり共産党本部も置かれ、地域の行政の中心を担っていた。でも現在は人影もまばらで5階建てほどの低層アパートが軒を連ねるメインストリートは閑散とした雰囲気だ。数軒ある食料品店の前にちらほら見える人影もほとんどが迷彩服を着たゾーン管理の関係者ばかりという状況では、普段着姿の我々がかえって異質な存在に見える。夜になると、事故後のライフライン確保のため町中に施設された太いパイプラインが薄暗い黄色のナトリウムランプに鈍い光を放ち、川崎あたりの工場に迷い込んだような錯覚になる。
　チェルノブイリインターインフォームはメインストリートを少し外れたところにある。事故後のゾーンを管轄するウクライナ非常事態省の出先機関だが、名前とは裏腹に建物は簡素なもので敷地の庭では野良猫たちがキモチ良さそうに昼寝をしている。職員はカーキ色の迷彩服を着て一見こわもてではあるが訪問者には気さくに応対してくれる。彼らはここに常駐している訳ではない。決められた期間でゾーン圏外に避難し、また一定の期間が過ぎるとこの地に赴任する。年間の許容被曝量を超えないための措置ではあるが、誰から見てもあまり気持ちのいい仕事ではなさそうだ。
「仕事があけたらキエフで遊ぶのさ。でも女はチェルノブイリが一番だ」
　副所長のユーリは流暢な英語を操りウオッカ

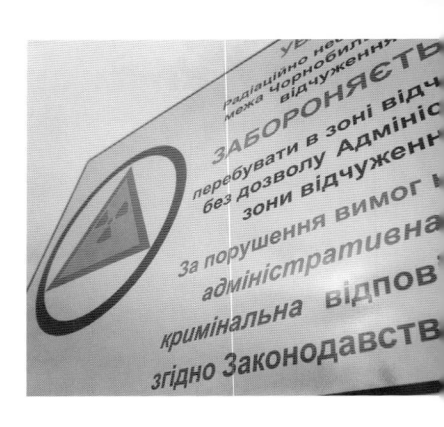

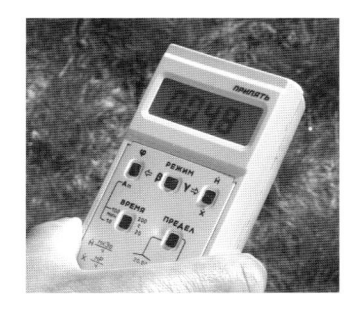

The air of spring is elegant everywhere. Aroma of beautiful flowers in various colors and the chorus of bush warblers and bulbuls carried by the soft spring wind reached us at the top of the 16-floored abandoned condominium. The gift of the spring was like a breeze of applause for us short of breath who had climbed up the dusty stairs breathing hard.

Pripyat is the town of Chernobyl Nuclear Power Plant. After the disaster, 500,000 residents left the town and it was the 23rd spring since then. On the top of the condominium, we clearly saw this ghost town, which was left in the big lowland called Polesia created by Dnieper River. The town spread on a radial to the north centering on the castle, the nuclear power plant. A few concrete condominiums were in the vast forest of broadleaf trees. The forest and the condos were like a desolate untended graveyard, cared for by nobody. But the greens of the trees were fresh and beautiful.

"This scenery is wonderful," Alexei, the interpreter, said. He has been to Pripyat for many years, but it was the first time for him to come to this place.

The area with the radius of 30 kilometers centering Chernobyl Nuclear Power Plant is called the "zone." The zone remains still off limits enclosed by a barbed wire fence due to the radiations emitted by the meltdown of the power plant 23 years ago. Although there are many places called ghost towns in the world, this town is unique because of its radiation contamination.

"Jun, please don't kneel down. I told you so for many times," Alexei said.

Because radiation hides itself like ninja, we cannot recognize its existence unless we detect it using a Geiger counter.

A small space can be a "hot spot" (a contaminated area of high-level radiation), Alexei said. He warned me that I could be exposed to radiation if I knelt down, a positional habit of a photographer.

"How about this tripod? It has already been stained with the earth," I asked him.

"Please dispose of it. It's a bad souvenir of Chernobyl," he said.

However, I tended to forget his repeated warnings as I was constantly amazed by beautiful fresh greens of the forests and the solemnity of the abandoned buildings that silently stood there. As said, radiation never overtly displays its existence to us.

"Let's finish reporting now and go back to the town," Alexei said. He warned me again by pointing out the Geiger counter showing "0.5 mili roentgen" when I was absorbed in taking pictures.

We let the spring scenery of the forests in Chernobyl, which spread like a vast green sea, imprint on our eyes. Then we ran down the rotted stairs of the building.

Chernobyl, small town

The city of Chernobyl is small town located a dozen kilometers south of the nuclear power plant. When we visited Chernobyl, it was right after the trees budded and the town was covered with white flowers of horse chestnut called the aesculus turbinate. While the town is smaller than Pripyat, it had a church and a headquarters of the Communist Party, and played the central role of the administration of the area before the disaster of the nuclear power plant.

But now the town is quiet and the main street with five-storied apartments aside is deserted. A few people who walked in front of some grocery shops were in camouflage, which meant they worked for the control of the "zone." In contrast, we were in normal clothes and looked like aliens. At night, a network of thick pipelines constructed to secure lifelines after the accident shined in the dark yellow light of sodium lamps. The scene gave me an illusion of getting lost in a huge factory.

Chernobyl Inter Inform, the office of the Emergency Situations Ministry of Ukraine managing the zone, was located a few blocks from the main street. While the name of the office sounds rigid, the building of the office was simple and stray cats were taking naps comfortably at the garden. The staff in camouflage looked tough at a glance. But they were friendly to visitors. The staff are not stationing here all the time of the year. After working at the building for a certain period, they evacuate outside of the zone and come back here again, so that they are not exposed to maximum radiation per year. Apparently the job is not pleasant one.

"After finishing my work here, I go to Kiev and have funs. But the best women in Russia are those from Chernobyl," Yuri, vice director of the office, said in fluent English while drinking vodka and pointing at Luda, the staffer of the office.

"That's right. I am Miss Chernobyl. You are lucky to see me," Luda said. Then she hugged me with her big body, typical of middle-aged Ukrainian women and tried to kiss me with her thick lip.

"Hey Jun, she is the souvenir from Chernobyl. Take her to Japan!" drunk guys said excitedly. They were shouting in joy and jeering in Ukrainian, English and Japanese, which created a chaotic atmosphere. In Ukraine, people cheered my appearance of bundled long hair when they were drunk, because the style seemingly looks like Ukraine warriors.

"Hey, samurai! Drink another glass of vodka," they said. "…I drink in my pride as samurai," I said. But the 10th glass of vodka burned my throat. "Vodka is the specific remedy of radiation sickness!" they claimed.

Small drinking party after dinner is modest pleasure because the town has no leisure except a little bar attached to a variety store. My reporting team and staff of the Inter Inform were gathering one by one and they brought various food and drink: vodkas, beers, cheese and dried river fishes. They drank, talked about silly things and laughed until midnight. "Why do you come all the way to such a place? If I were you, I would take pictures of pretty girls," Yuri told me at the peak of the excitement of the party.

"For me, this is a pretty place in a sense. Otherwise I wouldn't come here for some time," I answered.

"Pretty!? You are joking. What is pretty in the radiated ruins?"

"Listen. Ruins show us ourselves. Among the ruins, these here in Chernobyl show how human beings are. They become light and shade in the buildings…an infrastructure of pattern…"

"Well, I don't really understand. I don't like places where nobody is, because I feel lonely."

Ruins and Chernobyl

Abandoned buildings have an existence that people often find scary or avoid looking at. A friend of mine who has extrasensory perception warns me that I may be caught by negative power at ruins. Another friend, who is doctor, warned me that chemicals released from the abandoned buildings harm my health. But I find myself caught by the mysterious attractions of ruins ever since I played at war shelters as child. The sense of wandering in ruins may be similar to that of a fetus who is searching for the exit of womb unconsciously. When human beings are born, they find light and darkness are in contrast. The contrast occurs in close proximity; light and darkness next to each other. The truth can be understood when wandering at abandoned buildings. The sense printed on my genes developed in my childhood and accumulated, then led me to start visiting ruins when I grew up.

The town of Pripyat was the space where light and shade was crossing when I visited there for the first time in late autumn 2007. Under the clear blue sky, the leaves of poplars were shining in gold. However, once entering decaying rough concrete buildings abandoned some 20 years ago, I only found spaces that were dimly lit, leading to darkness.

After visiting a supermarket, a hotel and an amusement park in the center of the town, we made sandwiches of leftover breakfast and ate them with water. Sun shined but the air from northern plain of Russia was cold and heavy. It is unexpectedly cold in late autumn in Ukraine. Sasha, my guide, took out a bottle of vodka from his chest pocket and drank two caps of the liquor. Then he said, "Let's go to a movie theater now," and pointed out east. The theater of Pripyat was on the "riverbank street," located off the downtown. In this town, buildings with four stories or more are residence and three-storied-buildings or lower ones are hospitals, stores or other public facilities. I could imagine roughly what kind of facilities they were when I saw the appearance of the buildings. But before entering them actually, I could not know exactly what they were.

The exterior of the theater was well designed, which was different from simple and boring appearance of most architecture of the Soviet Union. The arched wall of the building was decorated with colorful mosaic tiles, though some of them had fallen off. The atmosphere was so warm that I almost forgot it was a ruin.

"Please be careful not to raise the dust," Alexei said when we walked through debris in the building and searched for the hall. Finally we found a thick door of the hall and opened it. Although I thought the inside of the hall was dark, it was just dim, the opaque curtain covering the ceiling window was ragged and the sun got in through it. We saw dusty chairs and tattered curtains. The slope of the floor was decayed and puddles of water had formed on the surface probably because of the leaky roof. We carefully walked down the slope and looked back at the whole hall from the side of the stage. It was big enough to accommodate 200 people. The interior was simple but somehow made me feel nostalgic.

"Let's go out from another door. It is dangerous to go back to the entrance," Sasha said, trying to open the door at the side of the stage. It was not locked but firmly closed. It might not have been opened for the last two

をいっきにあおって、宿守のリューダを笑いながら指差す。
「そうよ、私がミス・チェルノブイリ。私に会えたことはラッキーよ」
　というなりウクライナ人の中年女性の典型のようなその恰幅のよい躯で僕のことを抱きしめ、厚い唇を寄せてくる。
「ジュンサン、チェルノブイリのお土産だ。日本に連れて行っちまえ！」
　酔いも回った皆からは歓声と野次が飛び、ウクライナ語と英語と日本語が混じった混沌とした空間が生まれる。長髪を後ろでくくった僕の出で立ちがウクライナの戦士コサックに似ているらしく、酒の席ではやたらと受けがいい。
「おいサムライ！　もう1杯やれ」
「……サムライのメンツにかけていただきます」
　10杯目のウオッカはさすがに喉を焼く。
「ウオッカは放射能の特効薬さ！」
　雑貨屋の脇に小さなバー・スペースがある以外これといった息抜きの場がないこの街は、夕食後の小さな酒宴がささやかな楽しみとなる。撮影スタッフはもちろんインターインフォームの職員も三々五々集まり、いろんな種類のウオッカにビール、チーズや川魚の干物など皆それぞれ持ち寄って、他愛もない会話に笑いの絶えない宴は深夜まで続くのだ。
「なんで、わざわざこんなところにくるんだ？　俺だったら綺麗な女の子の写真撮るよ」
　宴もたけなわの頃、ユーリが赤ら顔で訊いてきた。
「僕にとってはある意味綺麗な場所だ。でなきゃ何度もこないよ」
「綺麗だって！？　とんでもない。放射能まみれの廃墟のどこが綺麗なんだよ」
「いいか、廃墟はね自分を映すんだよ。その中でもここはね、人類の姿を映すんだ。それが建物の中の光と陰になって……」
「うーん、ますますよくわかんねぇな。俺はとにかく人がいない場所は好きじゃないよ。だって淋しいじゃないか」

廃墟　そしてチェルノブイリ

　廃墟というものはとかく人々に不気味がられ時には煙たがられる存在でもある。霊感の強い友人はマイナスパワーに取り憑かれることを忠告し、医者の友人は建物の残骸から出る化学物質の人体への影響を忠告する。でもその不思議な魅力を子供の頃──防空壕で遊んだ頃──から感じている自分がいる。
　廃墟を彷徨うことは子宮の中で無意識に出口を求めている胎児の感覚に似ているかもしれない。暗黒と光明は両極でありながら実は隣り合わせであるという、人間が生まれてくる時に初めて悟る真理を廃墟を彷徨っているとたびたび感じることがある。僕が大人になって廃墟を巡り始めたのも、遺伝子に刷り込まれたその感覚が幼少時代の体験で増幅され蓄積されている結果に他ならないのだろう。
　プリピアチの街も光と陰が交錯する空間になっていた。初めて訪れた2007年晩秋。黄金色のポプラの木と秋の日差しが織りなすまばゆいばかりの外の光景とは一変して、崩れかけた武骨なコンクリートの建物に一歩踏み入るとそこは薄暗く闇へと続く空間が広がっていた。
　僕らは街の中心部にあるスーパーマーケット、ホテル、遊園地を巡ったあと、車の中で朝食の残りでサンドイッチもどきをこしらえて、水で胃袋に流し込んだ。太陽の光は降り注いでいるのだがはるか北のロシアの平原から垂れ込む空気は重く冷たい。ウクライナの晩秋は思いのほか冷えるのだ。ガイドのサーシャはウオッカを胸ポケットから取り出して瓶の蓋をグラス代わりに2杯ほどたて続けに飲み干しながら、
「今度は映画館に行こう」
　と東を指差した。プリピアチの映画館は中心街の外れの「川岸通り」沿いにある。この街では4階以上の高層の建物が住宅で、それ以下は病院、商店などなにがしかの公共施設である。大体の傾向はつかめて来たが、実際のところそれが何の施設なのかは入ってみないと分からない。だがさすがは映画館、ソ連建築の殺風景さはなく外観もしっかりとデザインされている。緩やかな弧を描く壁面には歯抜けではあるが色とりどりのタイルでモザイク画が施されて、廃墟であることを忘れてしまいそうなくらい暖かい雰囲気だ。
「あまり埃はたてないでくださいね」
　アレクセイの注意を背中に受けながら僕らは瓦礫をかき分け映画館を彷徨い分厚いホールの扉を開けた。暗いはずのホールは天井の明かり取りをふさぐ天幕が朽ち果てて、仄かな光が差し込んでいる。埃だらけの椅子や緞帳の残骸、雨水が天井から漏れてきたのかスロープの床は

腐敗し、水たまりができている場所もある。僕らは足場を確かめながら、おそるおそるスロープを下り舞台の脇からホール全体を振り返った。200人くらいは収容できそうな大きさで、何の飾り気もない質素な空間だが、どこか昭和の映画館を連想させる作りだった。
「別の場所から外に出ましょう。戻るのは危険です」
　サーシャは舞台脇のドアに手をかけた。ドアには鍵がかかっていないものの廃墟になって20数年、1度も開けられたことがないのかやたらと固く閉まっている。
「こういうのは俺に任せてくれ」
　僕はドアノブを両手で力一杯握り壁に足をかけて全体重を集中させ、一気にドアを引っ張った。ガリガリッと鈍い音とともに光が差し込み、舞い上がった埃がキラキラと輝く。
「こ、これは………」
　固く閉ざされたドアの向こうは何かの控え室のようだったのだが、その床は一面緑の絨毯になっている。床から繊毛のように伸びた苔には無数の細かい水滴がぶどうの実のように付着し、窓から差し込む光に反射して宝石をばらまいたかのようだ。そして目を凝らしてみると苔の隙間から1本のどす黒いキノコがひょろりと伸びている。僕は一瞬、屋久島の森に迷い込んだ錯覚に陥った。縄文杉に向かう途中の白谷雲水峡で見たあの苔が輝く原生林の光景だ。
　屋久島の原生林は生命の原始形態から極相状態までが混在しそれらが実に数千年のサイクルでゆっくりと推移しているという。生命を全うした巨大な杉の木は、自らが倒れることで辺りに光をもたらし、そして朽ち行くことで新しい生命のゆりかごとなる。倒木はやがて苔やキノコの菌糸に覆われて土となりそこに落ちた杉の種子が光を浴び育ってゆくというわけだ。
　文明の抜け殻とでも言うべき廃墟で、自然の営みは僕らのはかり知れない原始的なレベルでゆっくりと着実に進んでいるかのようだ。この映画館もあと数百年も経てば街路樹のポプラの子孫たちに埋め尽くされているのは間違いないだろう。チェルノブイリの廃墟は僕たちに放射能をまき散らした発電所事故といういわば文明の挫折という闇の部分を見せつつも、原始レベルでの自然の再生という一条の光を見せてくれる存在なのかもしれない。

赤い森

　チェルノブイリの街からゾーンの10キロ検問所を超えると道が二手に分かれる。左に行くとプリピアチに向かうショートカットで、発電所を右手に眺めつつ低木の原野の一本道に沿って車を走らせることになる。
「窓を閉めてくださいね」
　この道を通るときにアレクセイは必ずお決まりの車内アナウンスを行うのだ。一見すると何の変哲もない原野の一本道だが、今まで静寂を保っていたガイガーカウンターが赤ん坊の夜泣きのように突然けたたましく鳴き叫ぶ。液晶の数字は今まで見たことがなかった数字になった。
「ただいま.1.2ミリレントゲンです」
　この一帯は通称「赤い森」と呼ばれる。おとぎ話に出て来そうな名前ではあるが原子炉爆発後の火災で空気中に舞い上がった放射性降下物、つまり「死の灰」が大量に降り積もった場所だ。20,000レントゲンという人間の致死量を遥かに超えた放射線で、濃緑の松が赤く光ったとの目撃談から名もなき松林が「赤い森」と呼ばれるようになった。いまでも所々に枯れ木が立ち当時の様子を物語っている。
「車停めてもらってもいいか？　撮影したいんだが」
「えっ？　それは危険です‥‥でもどうしてもというなら‥‥えー、じゃあ2分だけ許可しましょう」
「2分だって!?」
　時間の根拠は曖昧だが、被曝の度合いというのは放射線の強さと放射線を浴びた時間をかけた量だ。1.2ミリなどたいしたことないようだが2時間浴びることで胸部レントゲン写真1枚分だという。
「草を踏んではいけません、汚染されます！」
　事故後に造られた道路は被曝をさけるため土が盛られ周囲より若干高いところを通っている。その土手を降りていこうとした僕をアレクセイは大声で制止した。そんなこと知ったものか。僕は制止を無視して土手を下り、アングルを探す。
「死にたいのですか？　死んでしまいますよ！」
　アレクセイの声は今までにないほどヒートアップした。それもそのはずだ。僕が下って

decades since it was abandoned.

"Let me open it," I said to him. I grabbed the doorknob with my both hands, pulled the door at a stretch by using my weight while bracing my foot against the wall. The door opened with sharp metallic sound, light and air entered at the same time, and we faced a fog of dust, it had been stirred up like a mist and it was shining in the air.

"What is this⋯"

The space behind the door was a stage room. But the floor was different; it was covered by a green carpet of moss. The mosses were growing; they looked like hairs of the carpet. Numerous drops of water were attached to the hairs; the drops seemed like grapes, little fruit that shined like jewelry in the light coming from the windows. When I peered into the room, I saw a long black mushroom growing among the moss. At that moment scale and location shifted, and it felt as if I was wandering in the forest of Yakushima Island, the island in southern Japan know for its thick primeval forests. I was walking through a ravine of one of the forests covered by shining moss, deep in the Shiratani Unsuikyo valley on the path that leads to a millennia old cedar called the Jomonsugi. Jomon is the hunting and gathering period prior to widespread agriculture. Sugi means cedar.

I have heard that the primeval forests in Yakushima Island are really strata, mixes of primeval states of lives and climaxes, which have slowly changed in the cycle of several thousands years. A huge cedar, which has lived for thousands years, reaches the end of its life, falls down and makes space for light. The fallen tree, and often the remains of its still upstanding trunk, is soon covered by a growth of moss and mushrooms, which consequently decay and change to soil. This earth becomes the cradle for a new round of life when seeds of cedars fall and grow, often inside the ruin of the tree trunk.

Such a cycle of nature at the primeval level seemed to be slowly and steadily proceeding forward at the abandoned theater in Chernobyl despite being something molted like a cast-off skin of civilization. With time, hundreds of years later, the theater would be covered by descendants of poplars on the streets outside whose seeds blew inside the remains. The ruins of Chernobyl show us the dark side of a failed civilization, the catastrophe of the nuclear power plant. But at the same time the ruins teach us hope through nature's recovery at a primeval level.

Red Forest

The road from the town of Chernobyl to the zone divided into two at the 10 km checkpoint from the nuclear power plant. The left road was the shortcut to Pripyat. Looking at the power plant on the right, we drove on the straight road in the field with low trees.

"Please close the windows," Alexei announces whenever we drive the road. It looked like a normal road, but Geiger counters which had kept silence suddenly started to make loud sounds like a baby crying at night. The number of the monitor exceeded 1 for the first time.

"Now it's 1.2 mili roentgen," Alexei said.

The area is called "Red Forest." Although the name is like something from a fairy tale, the place was covered by "death ashes", the radiated ash that flew from the fire after the explosion of the nuclear reactor.

The name came from witness accounts that the deep green pines of the forest shined in red when they were exposed to radiation that reached 20,000 roentgen, which was much higher than a fatal dose. Actually there were still dead standing trees as the result of the radiation exposure.

"Can you stop the car? I want to take pictures," I said.

"What? It's dangerous⋯But if you say you really want to⋯.. Well, I give you just two minutes."

"Two minutes?"

I was not sure the reason of the time length. But dose is calculated by multiplying the strength of radiation and the time of exposure. Probably going out for two hours there meant that I was exposed to radiation that equaled to having my chest X-rayed for a sheet of picture.

"Don't step on the grass! You are contaminated!"

The road constructed after the accident was made higher than the ground by landing up the earth to avoid radiation exposure. When I was trying to go down the mound, Alexei warned me in loud voice. But I ignored him and stepped down the mound to look for good angles.

"Do you want to die? You will be killed!"

Alexei's voice heated up to a high pitch, a level unreached before. I was the culprit responsible. The field I went down was the place where dead red pines exposed to radiation were cut down and buried. It was something like an underground storage of radiation. We cannot see, smell and sense radiation⋯. It is impossible to detect radiation with the five senses of humankind. Its existence is the fear that people can know only through knowledge. It is as if being attacked in a dark ring by enemies who existence can never be recognized.

When we went to a graveyard at the end of Pripyat, Geiger counters that could measure maximum 2 mili roentgen showed "error" because the radiation exceeded the upper limit. Apparently radiation still existed in the ground as if it hid itself there at random like land mines.

Some of the radiation sank in underground water and moved to other lands. Other radiation was absorbed by roots of plants and stayed in leaves or fruits. Radiation that reappears on the ground goes into the ecology and expands in many years. Actually periodical surveys of cesium 137 contamination areas showed long distant shifting of contaminated areas that specialists had never predicted. The shifts were not related to movements of wild animals or those of ecology. The radiation has been moving around invisibly at its own irregular pace even after 23 years of the accident.

The disaster occurred at Chernobyl Nuclear Power Plant at early morning on April 26, 1986. At reactor No. 4 at the plant, 200-tonage uranium fuels triggered nuclear fission, which ignited the graphite moderator that was surrounding the fuels. Inside of the reactor was full of various radioactive materials that had been generated by nuclear fission of uranium 235, main material of the fuels. About half of the uranium 235 became radioactive fall-out by the fire and were sent into the atmosphere through growing air current. The half-life of iodine 131 is eight days, relatively short, but it entered the grass, the cows which ate the grass, and the milk they produced. Eventually iodine was accumulated in thyroid glands of people who ate dairy products and a number of children suffered from thyroid gland disorders.

The half-life of Cesium 137 is about 30 years. Because the composition of the radiation is similar to that of potassium, vital mineral of animals and plants, cesium 137 is easily accumulated in the cells of the creatures. The half-life of strontium 90 is about 28 years. The composition is very similar to that of calcium and it is accumulated in bones of animals. Leukemia, one of major radiation sicknesses, is caused by strontium 90. The half-life of plutonium 239 is 24,000 years. The radiation is hardly discharged from the body and is accumulated in the lungs or kidneys, which releases strong α ray and destroys cells. Because of the long half-life, plutonium that spread in the not so long ago New Stone Age has its power now and is threatening life. It's like a black fairy tale; a tale composed of living ghosts and the lethal danger.

When our car passed the area of dead standing trees, the Geiger counters suddenly stopped making noise, like a baby fallen sleep. The scenery in front of us did not have anything that indicated the existence of radiation. There were plants and trees which colors change in accordance with the moving seasons and wild animals that inhabit the area on their own pace.

The driver stopped the engine of the car and we got out there. I sensed the aroma of spring flowers in the field, heard the songs of birds and found footprints of wild boars. What was missing here were the human beings, those who had left, leaving the land with concrete wreckage and radiation as a parting gift. Still the play of the seasons had come full circle. The part of the "Red Forest" seemed to change to "Green Forest", the full recovery of life achieved perfectly as far as we saw or could see with the naked eye.

But undoubtedly the radiation made by humankind had accumulated silently in lives with no words. I wondered whether the scene of such a rich abundant nature was the appearance of a fake recovery, a pseudo life masking sickness, or nature had the capability of accepting the awful power of radiation, receiving and transmuting it….

Visiting a village.

We were driving a highway while seeing chimneys of the nuclear power plant. But after crossing Pripyat River and going to a road, the chimneys were gone. The road was getting narrow and trees beside were growing their branches over the road. Our car lowered the speed and drove in the tunnel of fresh greens. The sound of the car surprised a family of wild boars and they crossed the road. Because horses were running at the grass land, which was former kolkhoz (collective farm in Soviet Union), and a nest of ibis was on the broken electric wires, I felt like going through a safari park. The paved road gradually changed to a road covered by ballast and sometimes by grass.

"We should have come on a jeep, right?" I said to Sergei, the driver.

"No, if the driving skill is good, it doesn't matter jeep or Renault," he said.

"You may be competing at the Dakar rally."

Sergei stroke a victory pose and dashed his new Renault into the wild road. Branches hit the frond glass of the car and made rubbing and creaking sounds on the ceiling. But he

いった場所は被曝し赤く枯れた松を伐採して地中に埋めたところだった。いわば放射能の地下貯蔵庫である。見えない、臭わない、感じない……。人間の五感では感知できずただ知識からくる恐怖として存在する放射能。暗黒のリングで気配のしない敵に攻撃を受けているといったところだろうか。

プリピァチの街外れにある墓地では2ミリレントゲンが計測限界のガイガーカウンターの液晶表示にエラーが出てしまった。放射能は依然として地雷のごとく地中にランダムに眠っていることは確かなようだ。そしてあるものは地下水へ浸透し別の土地に移動し、またあるものは植物の根に吸収され葉や木の実となって再び地上に姿を現す。地上に再び現れた放射能は生態系に潜り込み長い年月をかけて拡散する。事実事故発生後定期的に行われているセシウム137の汚染分布調査では専門の学者でも想像のつかない汚染の地域移動が確認されている。野生動物の移動や植生分布の移動と連動することなく、放射能は我々の不可視なところで独自の不規則な移動を今も行っている。

1986年4月26日未明に発生したチェルノブイリ事故では4号炉に挿入されていた200トンのウラニウム核燃料が連鎖的に核分裂反応をおこし、減速材として核燃料の周囲を覆っていた黒鉛に引火。原子炉内には燃料の母体であるウラン235の核分裂で生成された様々な放射性物質が充満しており、その約半分が火災の上昇気流に乗って塵埃となって大気中に放出された。ヨウ素131は半減期が8日と短いものの、牧草→牛→牛乳という過程を経て人体へ侵入し甲状腺に蓄積。多くの子供たちに甲状腺障害を引き起こした。セシウム137は半減期が約30年。物質の組成が動植物の必須ミネラルであるカリウムと酷似しており細胞内に蓄積されやすい厄介な物質だ。ストロンチウム90は半減期が約28年。組成はカルシウムと酷似しており主に動物の骨に蓄積される。放射線障害のひとつに白血病があげられるのはストロンチウム90の仕業である。プルトニウム239は半減期が2万4000年。体外に排出されにくく、肺や腎臓に沈着して強烈なα線を発し細胞を破壊する。例えれば新石器時代にばらまかれたプルトニウムが今も勢力を温存し牙を剥いているというブラックなおとぎ話になるのだ。

赤松の枯れ木が林立する一帯を通り過ぎるとガイガーカウンターは眠りについた赤ん坊のように急に静かになった。目の前に広がる光景には放射能を認識させるものは存在しない。季節の移ろいに従って衣をかえる草木があり、自らのサイクルで大地に生きる野生の動物たちがいる。車のエンジンを切って外に出ると、春の野花の香りが漂い、小鳥たちの歌声が響き、イノシシの足跡が見える。ただいないのはコンクリートの残骸と放射能という置き土産を残してこの地を去っていった人間だけなのだ。「赤い森」もいまでは「緑の森」に姿を変え、僕らの見える範囲では見事に生命の再生を勝ち取ったかのようである。しかしもの言わぬかれらの生命体の中には、依然として人間が作り出した放射能が蓄積していることは間違いない。眼前に広がる豊穣な自然の光景は見せかけの再生の姿なのだろうか、それとも自然には放射能の魔力をも受け止めてしまう不思議な底力があるのだろうか……。

村を訪ねる

幹線道路から見えていた発電所の煙突はプリピァチ川を渡り脇道にそれると視界から消えてしまった。道は次第に細くなり両脇の木々は道上まで枝を伸ばしている。僕らは新緑のトンネルの中スピードを落としながら走った。車の音に驚いたイノシシの大家族が道を横切ったり、ただの草地となったコルホーズ（ソ連時代の集団農場）を野生馬が走っていたり、折れた電線の上にコウノトリの巣があったりと、ちょっとしたサファリパークのようだ。舗装道路はやがて砂利道になり、場所によっては草に覆われて道なき道となる。
「ジープの方がよかったんじゃない？」
「いやいや、腕がよけりゃジープもルノーもおんなじさ」
「パリダカールに出られるかもな」

ドライバーのセルゲイはガッツポーズをしながら新車のルノーで獣道に突っ込んでいく。小枝がフロントガラスにバシバシと当たり、キリキリと嫌な摩擦音を天井で響かせているにもかかわらず障害を巧みにかわしながらしばらく走ると、やがて視界が広がりぽつりぽつりと茅葺屋根の廃屋が見え始めた。
「クラスノエ、日本語で『美しい』という村です」
チェルノブイリ事故で住処を追われたのはプ

リピァチの住人だけではない。ウクライナ北部、ベラルーシ南部の低地帯はこれといった開発がされることもなく以前からきわめて中世的で半自給的な村落が数多く存在していた。丸太を組んだ茅葺きの小さな家。中にはオーブンとヒーターと電気毛布？を兼ねるペチカがあって、牛やヤギを飼い菜園を耕し、人々は自然のサイクルに身を委ね小さな牧歌的暮らしをして来たそうだ。彼らは発電所で起こったことも空から降って来た放射能のことも、詳しく聞かされることもなく、突如現れた旧ソ連軍に住処を追われ家を焼かれて埋められてしまった。運良く旧ソ連軍の狼藉を免れた家も、住人が帰ってくることはなく、20数年の歳月を経てゆっくり自然に帰ろうとしている。一説では10万人もの人が長年住み慣れた土地を離れて行かざるを得なかったといわれている。

車はやがて村の教会の前で停まった。直線距離にして発電所から6キロほどしか離れていないのだが、こじんまりした素朴な村の残骸だ。核を操る高度な文明と小さな中世的な暮らしが隣り合わせだったことを目の当たりにする。僕らは外壁のペンキが剥げかかった教会の扉を開けた。がらんとした空間に床の軋む音が増幅される。天井のイエスと天使の壁画は色あせずに残っていたものの、祭壇に供えられた花は既に干涸びている。主イエスはこの村で起こった悲劇をどのように見ているのだろうか。

村の家の大半は丸太組みの茅葺き屋根で日本の農村と似た風情だが、さすがに20年以上も放置されていることもあって、屋根の茅は腐食し、壁面の木組みだけが残る家屋も目立つ。歪んだ窓から中をのぞくと青く塗られたペチカのペンキは剥げ、壁にかかった夫婦のコートにはうっすらとコケがはえ、水場に並んだ食器は埃で輝きを失っている。まるで神隠しにあったかのようにこの家の主は姿を消し、主を待ちくたびれた家は徐々に姿を変えつつある。幾年月も経てば「美しい」村もやがて「美しい」森へと姿を変えるのだろう。あまりにも無情な「再生」へのプロセスだ。

僕らはクラスノエ村をあとにして、チェルノブイリ市の雑貨屋で手みやげのパンや日用生活雑貨を買い、サモショールの老人宅を訪問するため車を東へと走らせた。サモショールとはロシア語で「居残り者」を意味するが、チェルノブイリゾーンでは「不法滞在者」という扱いがなされている。事故発生後の強制退去命令を無視して住み慣れた土地に居続ける人たちのことだ。今現在そのほとんどが70歳を超えた老人たちだ。

警察の検問所を通過して15分ほど走るとパリシェフ村に到着した。道路脇に小さなバス停が現れセルゲイは車を停めた。廃屋が連なる一角に煙突から薄い煙の出た小さな家が見える。電話も通じていない老人宅への突然の訪問だ。顔見知りのセルゲイは小走りで民家の裏手に回り、しばらくして我々を招き寄せた。垣根伝いに歩いてゆくと、ニワトリやアヒルや猫たちが突然の訪問者に驚き雲の子を散らすように茂みに身を隠す。

「プルィヴィート（こんにちは）！」

76歳のマリアババーシカ（マリアおばあさん）が歯の抜けた笑顔で我々を家に招き入れた。どこの国でも老人は元気だ。アレクセイが通訳する隙間を与えないほどの早口でしゃべり、その間も不自由な足を引きずって突然の無礼な来客をもてなす準備をしている。くすんだ赤いクロスを掛けた小さなテーブルにはウクライナの餃子ワレニケをはじめチーズやパン、ウオッカグラスも並べられた。

「ヤポン（日本人）はみんな髪が長いのか？」

ババーシカは僕の長髪を指差し、

「ウクライナのコザックも髪が長い。コザックもヤポンもロシアと戦った。だから友達だ」

としゃべりながら奥の部屋から大きなガラス瓶に入った透明の液体を抱えながら持ってくる。セルゲイは僕に目配せをし、マッチをする仕草をする。

「ババーシカのウオッカは特別だ。寒い冬もペチカとコイツがあれば大丈夫」

セルゲイは瓶からウオッカを少し取り床に撒きライターの火を近づけた。するとどうだ、床には形の定まらない青白い炎がゆらゆらとたなびく。セルゲイは炎を靴底でもみ消しながらガッツポーズ。

「純サンこれが飲めるなんて幸せだよ」

とにやりと笑みを浮かべる。ババーシカはウオッカグラスに自慢のウオッカを注ぐが、僕の前に置かれたグラスは日本でいうビールグラスの大きさだ。

「ブギモン（乾杯）！」

ウオッカは一気に飲み干すのがたしなみだが、日本では酒豪のはずの僕でもビールグラスになみなみとつがれたまさしく火の出るウオッカを一気に飲み干すのはつらい。3分の1も飲

managed to drive through the obstacles for a while and we could see some abandoned houses with thatched roof in open field.

"This is a village named Krasnoe, meaning 'beautiful,'" Alexei said.

The disaster of Chernobyl Nuclear Power Plant forced residents of Pripyat to leave their town, but also residents of villages at lowlands in northern Ukraine and southern Belarus. The areas had not been developed and people had lived in semi self-sufficient farms in medieval life style in some villages. In the small houses made of logs and thatched roofs, people used Russian stoves called pechka for cooking and heating. I heard that the villagers had lived modest and pastoral lives in the natural cycle by keeping cows and goats and raising vegetables. However, right after the nuclear power plant disaster, the army of former Soviet Union suddenly came into the villages. They drove the villagers out of the houses, burned the houses and buried them. The villagers had never been explained enough about the accident and radiations that fell on them from the sky. Even though some houses were not burned luckily, the residents of the houses never returned and they were slowly decaying and becoming the part of the nature in the two decades. It is said that about 100,000 people were forced to leave their lands where they had lived for many years.

Our car stopped in front of a church in the village, which was only six kilometers far from the nuclear power plant. It was the wreckage of a small and simple village. It showed clearly the fact the highly developed civilization that operated nuclear power and a small medieval life had been next to each other. We opened the door of the church with exterior walls which paint was falling off partly. The squeaking sound of the floor echoed loud in the empty space. Although the painting of Jesus Christ and angels on the ceiling haven't been faded, flowers on the altar dried up. I wondered how the lord, Jesus, has seen the tragedy happened in this village.

The majority of the houses were made of logs and thatched roofs, which looked similar to the appearance of agricultural villages in Japan. But because those in the village in Ukraine were left without care for more than two decades, the thatches decayed and many of the houses stood with only wooden pillars on the walls. When I looked into one of the houses through distorted windows, blue paint on the pechka was falling off, mosses were growing on the coats of a husband and a wife hooked on the wall, and dusty table wares at the kitchen lost their shiny outlook. The owners of the house disappeared as if they were spirited away. The house which had waited for its owners coming back were changing itself gradually. When several more years passed, the "beautiful" village would change to 'beautiful' forest. It's quite merciless process of "recovery."

Leaving Krasnoe Village, we went back to the city of Chernobyl and bought bread and sundry goods as souvenir for senior people living in west. They are called "illegal residents" in Chernobyl zone. They have ignored the expulsion order after the accident and stayed in the land where they had got used to live. Most of them are the senior whose age is over 70 years.

Our car passed the checkpoint of the police, ran for 15 minutes and arrived at Parishev Village. Sergei stopped the car at the roadside where a small bus stop was. Among a series of abandoned houses, we could see a small house with a chimney blowing off thin smoke. Because the house didn't have telephone line, we were trying to visit the old woman lived there without any notice. Sergei, who knew the old woman well, ran to the backyard of the small house and then called us. When we walked through the hedge, hens, ducks and cats were surprised by these sudden visitors and dashed into the bush to hide themselves.

"Hello!"

"Hello!" said the 76-year-old Maria babushka, whose teeth fallen out, smiling and inviting us to her house. We can find lively old people like her in any country. She talked so fast that Alexei could not find time to interpret and prepared for entertaining us, the sudden visitors, while walking with her disabled leg. She placed Ukrainian pork dumplings, cheese, bread and glass of vodka on the small table with dull red table cloth.

"Do all Japanese men have long hair?" the old lady asked, pointing out my long hair.

"Cossacks of Ukraine had long hairs, too. Cossacks, as well as Japanese, fought Russians. So we are friends," she said, bringing a big bottle of transparent liquid from another room.

Sergei winked me and showed a gesture of firing with match.

"Vodka of babushka is special. You can survive cold winter here by warming yourself with this and pechka," he said.

Sergei poured a bit of vodka from the bottle on the floor and brought a fire of his lighter to the floor. Then, amazingly, a blue white fire in unclear form appeared and trailed. Sergei smothered the fire with his shoe sole and

stroke a victory pose.
"Jun, you are so lucky to drink this," he said and grinned.
Babushka poured her special vodka into glasses but the one in front of me was big enough to be called beer glass.
"Cheers!"
In Ukraine, it is a manner to drink up vodka. But it was hard even for me, heavy drinker in Japan, to drink up the beer glass full of the vodka that could generate fire. When I drank one-thirds of the glass and put it on the table, babushka shouted something.
「○△▽＊＋¥■」
"What did she say?"
「¥■%&＄＃？＊〜！」
"Jun, babushka says you should drink up the glass," Alexei, who doesn't like vodka much and was at a loss, said while gesturing emptying the glass.
"I am so honored to see Maria babushka!"
While everyone was looking at me, I took the glass, opened my mouth wide and poured the rest of the vodka in one. My gullet and stomach became hot rapidly.
"Good job, Japanese! Listen to my Cossack song," babushka said and sang her favorite song in sexy and beautiful voice that let us forget her age. She even danced elegantly while singing though she was stepping with her disabled leg.

Burning pines
Burning
The girl under the tree
Tied her hair on the shoulder into braids
And mumbled
Oh, my braids, braids
Thank you for long time
I must say good-by soon
You are going to be covered by a white handkerchief when I marry
Burning pines
Over the smoke is a man
Oh, my love

We applauded babushka and she smiled. She looked like a girl. I wondered how was the life of babushka who had lived in the land since her she had been born and loved the land very much. I was moved by her toughness as human being. She was as powerful as the plants and trees that grew their buds from the contaminated land.
"I can harvest a lot of mushrooms here and vodka is delicious. This is the best place to live!" she said and laughed by opening her mouth with fallen teeth.

Another Maria

In the morning of the final day of my previous trip to Chernobyl, Sergei, the driver, became a grandfather. His daughter, Nataliya, gave birth to a baby girl at a hospital in Kiev. He was humming while driving and smiled all the time.
"She must be a beauty because she is my granddaughter. She will be ballet dancer or actress when she grows up," he said.
"Then I will be photographer for her," I said.
We were talking so cheerfully that we forgot the fact that we were driving in Chernobyl.
Later in Japan I learned that the baby was named Maria. One of the reasons I wanted to visit Chernobyl in spring was to see Maria, who is growing, with my own eyes.
The house of Sergei was at a small village located a bit south of the 30 km checking point from the zone. The pretty house with ivy on the walls had a big garden covered by various flowers. It had a kitchen garden and seedlings of summer vegetables had been planted. We were guided to a small detached room that Sergei had built by himself.
"Hey, Maria! It's me, your grandpa!"
Hearing her grandfather's voice, little Maria toddled and appeared at the entrance to welcome us. But seeing aliens waving their hands, Maria dashed back to her mother and buried her face in the chest of her mother. The little girl sometimes gave quick glances at us as if she wanted to see something scary. She was actually a Ukrainian beauty who looked like her mother, just as Sergei had expected.
"Hello, Maria! I was looking forward to seeing you!" I smiled, slowly approached her and tried to give her a doll, a souvenir for her. However, her big blue eyes were soon filled with tears. After a silence, she began to cry out loud. Nataliya cradled her and Sergei looked at them with gentle eyes. Three generations together. It was a daily scene of a common Ukrainian family.

- - -

The contaminated area of Chernobyl is the paradise created and lost by human beings. Radiation was released to damage all lives and a number of precious human lives were lost. People were no more allowed to live there. Despite the fear from humankind, plants and trees started growing to recover their primeval state gradually. It seemed a primitive polesia (low land) revived as the cast-off skin of the highly developed civilization.
Nataliya, Sergei's daughter, was born at this

むと喉が焼ける。仕方なくグラスをテーブルに置くとバブーシカが叫ぶ、
「○△▽○＊＋¥■」
「えっ、なんだって？」
「¥■％＆＄＃？＊〜！」
「……純サン、バブーシカが飲み干せと言っております」
ウオッカがあまり得意ではないアレクセイが困惑しながらグラスをあける仕草をする。
「マリアバブーシカに逢えて光栄です！」
皆の視線が集まる中、僕はグラスを手に取り口元に近づけ喉を大きく開いて一気に残りのウオッカを流し込んだ。喉が焼け、食道も胃も一気に熱を帯びる。

「よくやったヤポン！　私のコザックの歌を聴いてくれ」
バブーシカは足を引きずりながらも軽やかな振りをつけて、歳を感じさせない艶やかな声で自慢の唄を披露してくれた。

松が燃える
燃えさかる
木の下にいる女の子
肩にかかったみつあみを
そろえながらつぶやいた
ああ、みつあみよ、あたしのみつあみよ
長い間ありがとう
もうすぐお別れだね
角かくしの白いハンカチで
すっぽりおおわれてしまうのだから
松が燃え
煙の向こうに一人の男
ああ、あたしの愛する人よ

拍手喝采のもとバブーシカはにっこり笑う。その笑顔はまるで少女のようだ。生まれてこのかたこの土地に住み続けこの土地を愛してやまない彼女の生き様はどんなものだったのか……。僕は汚染された大地から再び芽を出す草木たちに負けず劣らずの人間の強さを感じた。
「キノコはたくさんとれるし、ウオッカもうまいし最高の場所だよここは！」
バブーシカは歯の抜けた口を大きく開いて高らかに笑った。

もう一人のマリア

前回のチェルノブイリ取材旅行の最後の朝、ドライバーのセルゲイがおじいさんになった。娘のナターシャがかわいい女の子を産んだとキエフの病院から連絡があり、運転中も鼻歌まじりで顔がほころんでくしゃくしゃだった。
「僕の孫だから絶対に美人だ。将来はバレリーナか女優だよ」
「じゃ僕は彼女の専属カメラマンってことでどうかな……」
車内はここがチェルノブイリであることを忘れるほど陽気な話に花が咲いた。
後に日本でその娘がマリアと名付けられたことを知った。僕が春のチェルノブイリを見たいと思ったひとつの理由は、マリアが成長している姿をこの目で確かめたかったからである。
セルゲイの家は30キロゾーンの検問所をわずかに南に下った小さな村にある。壁に蔦をはやした瀟洒な家で、広い庭先は百花繚乱、家庭菜園もあって夏野菜の苗が既に植え付けられている。彼が娘夫婦のために自力で建てたという小さな離れに案内された。
「おーい、マリア！　おじいちゃんだよ！」
祖父の声に反応してよちよち歩きで現れたマリアは玄関先で我々を迎えてくれたが、初めて見る異邦人が手を振る姿を見るや身を翻し、母親の胸に飛び込み顔をうずめた。恐いもの見たさか時折ちらちらと視線を送ってくるが、セルゲイの予想した通り母親似のウクライナ美人であることは間違いなさそうだ。
「マリア、こんにちは！　会いたかったよ！」
僕は微笑みながらゆっくり彼女に近づき、お土産の人形を差し出した。途端に彼女のつぶらな瞳は涙で満ちあふれ、しばしの沈黙のあと大きな泣き声が離れの中に響き渡った。それをあやす母親のナターシャ、そしてその姿を優しい目線で見守るセルゲイ。ウクライナの普通の家庭の普通の光景が僕の目の前に展開していた。

＊　＊　＊

チェルノブイリの汚染地帯は人間が自ら作り出してしまった失楽園である。
すべての生命に打撃を与える放射能がばらまかれ、多くの尊い人命が失われ、住むことさえも許されなくなった。ただ人間が恐れおののく

のを尻目に、草木は原始の成長の姿を徐々に取り戻した。高度な文明の抜け殻には果てしなき原始のポレーシェが蘇ったかのように見える。

ナターシャは発電所事故が発生してほとぼりも冷めないうちに失楽園から少し離れたこの村で生まれた。今まで常に放射能の危険や恐怖と隣り合わせで生きて来たのは間違いのないことであろう。でもその21年後新しい命、マリアがこの世に生を授かった。このことは奇跡なのかそれとも人間本来が持っている逞しさなのか？ 汚染された大地の傍らで人間もまた自然の流れに沿った生命のサイクルを繰り返している姿を見て、少し安心した気分になった。

春のチェルノブイリは淡い青空の下若い緑が目を和ませてくれた。季節が移ろい一年が過ぎても、少し背たけが高くなったポプラの木と子孫を増やした路傍のタンポポの花が、再び僕を迎えてくれるにちがいない。そしてマリアばあさんのウオッカには磨きがかかり、少女マリアは駆けっこができるようになっているだろう。

チェルノブイリの自然と人間が織りなす人知れぬ小さな時間の移ろいに、僕らの未来へのちいさなメッセージを見たような気がした。

耕す　〜あとがきにかえて

念願かなって近所に畑を借りて2年ほど経つ。わずか100坪とはいえたまに友人の手を借りる以外はほとんど自力で土と戯れるのは思った以上の重労働だ。そして何より加減が難しい。肥は入れすぎても駄目だし、雑草も根絶やしにしてはいけない。収穫を狙っている獣や鳥にも分け前を上げたいのだが、ほおっておくと自分の分け前がなくなる。日々思索の連続だ。

僕が土を耕し始めたのはすべての命の再生の源になる「光」や「水」や「土」に対する畏敬の念からだ。放射能汚染されたチェルノブイリで逞しくも四季のサイクルを循環させている植物の秘めたる強さに心を打たれたからだ。無心になって鍬をふっていると、様々な思いが脳裏を去来する。

アボリジニを研究している大学の恩師がゼミで言った言葉がある。

「原発の賛成派も反対派も表面にかかげるスローガンには間違いがある。原発はけして安全ではないし、耳かき1杯のプルトニウムで万単位の人が死ぬことはない。情報化社会になればなるほど本当に知りたいことは見えなくなるものだ。皆さんは真実を突き止める力を磨いてください」

真実とはいったいなんなのか。「核」の暴走が巻き起こす現実は、チェルノブイリの原野と置き去りにされた文明の抜け殻が一番よく語っていることだ。

僕のできることは微力ながらありのままの姿を伝えることであるし、本書を手に取ってくれた読者の皆さんにはそれに対して感じてもらうことだろう。文明の恩恵を享受しているという矛盾を意識しつつ、固く締まった世の中の趨勢という痩せた土地に皆で孤高の小さな一鍬を振り下ろせれば幸いだと思っている。

最後に本書の出版に尽力してくださった二見書房（株）の米田郷之氏、素敵な装丁をしてくださったヤマシタツトム氏、無理なスケジュールで翻訳を受けてくださった有田えり子氏、原稿執筆の資料探しに協力してくださった小野元浩氏、現地撮影をサポートしてくれたアレクセイ氏、サーシャ氏、セルゲイ氏、そして本書を手に取ってくれた読者の皆様に感謝いたします。

チェルノブイリ事故25年の春

中筋　純

追記：偶然にも本書製作中の2011年3月11日に東日本太平洋沖地震が発生し、安全といわれていた福島第1原発が日本の原発史上最悪の事故を起こす結果となりました。状況の推移を注視しておりましたが、こと情報の公開に至っては25年前のチェルノブイリの教訓が少しも活かされていないことにいらだちを感じました。最後に被災者の方々のご心中を察するとともに深くお見舞い申し上げます。（2011年3月16日　八王子の暗室にて）

village not far from the paradise lost after the disaster of the nuclear power plant. There is no doubt that she has lived for more than the two decades up until today always being aware of the danger and the fear of radiation. But after 21 years, she was given a new life, a daughter Maria. Is this a miracle or the toughness of human beings living the forces of nature? I felt a little bit relieved as I saw not only nature but also human beings were repeating their natural life cycle even though they live in the land close to the contaminated area.

In Chernobyl in spring, fresh young greens under a pale blue sky pleased my eyes. When the seasons change and more years pass, taller poplars and dandelions, which will have increased their own children on the roadside, will welcome me again. The vodka of grandma Maria is going to be more delicious and other little Marias will be able to run about. Nature and humankind shall weave changes in Chernobyl as time flows, though few people in other parts of the world will know about them. I thought I saw a small message for our future in the changes.

Cultivating ~Afterword

It has been two years since I rented a farm, which I had wanted, in my neighborhood. Although it is only 330 square meters big, it takes much more effort and labor than I expected to cultivate the farm by myself except when friends of mine help me once in a while. It is the most difficult to adjust things in the farm. It is not good to put too much fertilizer and to pull out all the weeds. I wish I can give some of the harvest to animals and birds. But if I leave it to them, my share is gone. I am struggling everyday.

I began to cultivate land because I was awed by "sunshine," "water" and "earth". These are the source of recovery of all lives. I was moved by the secret strength of plants that circulates the changes of the seasons in the land of Chernobyl contaminated by radiation. When I am cultivating the land in my neighborhood by using hoe, a lot of thoughts come to mind.

I remember words of my mentor in the university who has studied aboriginal people in Australia.

He said in his seminar; "There are mistakes in the slogans used by both pro-nuclear power groups and those against nuclear power. Nuclear power plants are never safe and milligram of plutonium cannot kill tens of thousands people. There are doomsday and non-doomsday scenarios. It is getting more difficult to learn what you really want to know as the society is becoming increasingly information-driven. Information is not truth. Improve your capability to find out truths."

What are the truths? The reality of the pulled trigger and consequent runaway of nuclear power is the most obvious in the fields and the abandoned civilization in Chernobyl.

What I can do is to try to inform readers of the whole truth and to think about it but my ability is not enough. I am aware of the contradiction that I enjoy the benefits of civilization. But I hope we can make small changes in the conservative trends of the societies, maybe people cultivating poor lands using hoes by themselves can be something.

Lastly I would like to express my gratitude to Satoshi Yoneda of Futami Shobo Publishing Co. for his great efforts in publishing this book, Tsutomu Yamashita who made nice binding, Eriko Arita who translated the manuscript in tight schedule, Motohiro Ono who cooperated in researching materials for my writing. I also appreciate Alexei, Sasha and Sergei who supported my reporting in Chernobyl and people who read this book.

In the 25th spring since the disaster in Chernobyl,
Jun Nakasuji

Postscript: By accident when I was compiling this book, catastrophic earthquake hit northeast of Japan and caused a series of accidents at Fukushima No. 1 nuclear power

plant. They are the worst-ever disasters in the history of nuclear power in this country. While watching the changing situations at the plant, I was frustrated by the fact that lessons of Chernobyl 25 years ago were not utilized at all by the authorities especially in providing accurate information for the public. Lastly I would like to express my heartfelt sympathy to the people who suffer from the
earthquake.

At the darkroom of my studio in Tokyo's Hachioji
On March 16, 2011

中筋　純　Nakasuji Jun

1966年、和歌山県生まれ。東京外国語大学中国語学科在学中にアジア、中米を放浪。卒業後出版社に勤務しつつ独学で写真技術を習得する。1996年に独立し中筋写真事務所設立。数々の雑誌をメインにアパレル広告、舞台広告、CDジャケット撮影など多岐にわたって活躍中。著書に『廃墟チェルノブイリ』（二見書房）、共著に『廃墟本』シリーズ（ミリオン出版）、『廃墟、その光と陰』（東邦出版）などがある。2009年個展「黙示録チェルノブイリ」（キヤノンギャラリー）、2011年個展「黙示録チェルノブイリ　再生の春」（ニコンサロン）開催。
E-mail:suzy_j1966@yahoo.co.jp

ブックデザイン	ヤマシタツトム
翻訳	有田えり子
現地撮影コーディネイト	アレクセイ・タナシエンコ http://www.japanese-page.kiev.ua
参考文献	『チェルノブイリの森』（メアリー　マイシオ著　中尾ゆかり訳） 『チェルノブイリ　消えた458の村』（広河隆一編著） 『チェルノブイリ　極秘』（アラ　ヤロシンスカヤ著　和田あき子訳） 『チェルノブイリの放射能と日本』（寺島東洋三／市川龍資編著）

チェルノブイリ　春

写真・文　中筋　純

発行所　株式会社二見書房
東京都千代田区三崎町2-18-11
電話　03-3515-2311（営業）
　　　03-3515-2313（編集）
振替　00170-4-2639

印刷・製本　図書印刷株式会社

乱丁・落丁本はお取り替えいたします。定価はカバーに表示してあります。

© Nakasuji Jun 2011, Printed In Japan.
ISBN978-4-576-11055-4
http://www.futami.co.jp